Modeling Self-Heating Effects in Nanoscale Devices

Modeling Self-Heating Effects in Nanoscale Devices

K Raleva
FEIT- Sts Cyril and Methodius University, Skopje, Republic of Macedonia

A R Shaik
Arizona State University, Tempe, AZ, 85287-5706, USA

D Vasileska
Arizona State University, Tempe, AZ, 85287-5706, USA

S M Goodnick
Arizona State University, Tempe, AZ, 85287-5706, USA

Morgan & Claypool Publishers

Rights & Permissions
To obtain permission to re-use copyrighted material from Morgan & Claypool Publishers, please contact info@morganclaypool.com.

ISBN   978-1-6817-4123-9 (ebook)
ISBN   978-1-6817-4059-1 (print)
ISBN   978-1-6817-4251-9 (mobi)

DOI   10.1088/978-1-6817-4123-9

Version: 20170801

IOP Concise Physics
ISSN 2053-2571 (online)
ISSN 2054-7307 (print)

A Morgan & Claypool publication as part of IOP Concise Physics
Published by Morgan & Claypool Publishers, 1210 Fifth Avenue, Suite 250, San Rafael, CA, 94901, USA

IOP Publishing, Temple Circus, Temple Way, Bristol BS1 6HG, UK

To Martin, Emili and Zdravko

Contents

Preface

For the last half-century, miniaturization of integrated circuits through transistor scaling has been the driving force for the semiconductor technology roadmap. The idea of a technology roadmap for semiconductors can be traced back to a paper by Gordon Moore published in 1965, in which he predicted that the number of transistors in a semiconductor chip will double every 18 to 24 months. This consequently results in a reduction of the cost per elementary function (cost per bit for memory devices, or cost per MIPS for computing devices), thus enabling the production of more complex circuits on a single semiconductor substrate. This trend towards miniaturization is known as 'Moore's Law'.

For several decades, the ability of the semiconductor industry to follow Moore's law has been the engine of an incredible cycle: through transistor scaling one obtains a better performance-to-cost ratio for integrated circuits (ICs), which has resulted in an exponential growth of the semiconductor market. This, in turn, allows further investments in new technologies which enable further scaling. Scaling in turn, enables creation of more complex, faster, cheaper and low-power ICs.

Moore's Law for CMOS (complementary metal–oxide–semiconductor) scaling has already hit the power wall. Even with new approaches for the technology roadmap, known as 'more-than-Moore' scaling, which involves parallelism and improving efficiency at all technology levels—architecture, software and devices, thermal-power challenges and increasingly expensive energy demands still pose threats to the rate of increase in computer performance. Many of the transistors must be powered-down at any given time to meet the power budget. Also, even though scaling allows a decrease in the power per transistor, the number of transistors on a chip is increasing faster than the power requirements are falling. During the first decades of Moore's law, constant field scaling ensured that when the device dimensions (along with supply voltages) were decreased by a factor F, the power density per unit area remained constant with increasing integration density. However, this approach had already broken down by the late 1980s as oxide thickness became too small, hot electron and short channel effects resulted in decreasing device performance with scaling, and power supply scaling slowed down considerably, resulting in an increase in power density over successive generations. As a result, clock speed has essentially remained constant over the past decade (since power dissipation is proportional to clock speed in CMOS). Increasing power dissipation and decreasing device dimensions also exacerbates device self-heating effects, causing an overall degradation of the on-current due to the characteristic phonon hot spot near the drain end of the device, which does not scale proportionally.

It is generally acknowledged that modeling and simulation are preferred alternatives to trial and error approaches to semiconductor fabrication in the present environment, where the cost of process runs and associated mask sets is increasing exponentially with successive technology nodes. Hence, accurate physical device simulation tools are essential to accurately predict device and circuit performance and to optimize design. At the nanometer scale, there are several effects that must be accurately captured in state-of-the-art device simulation tools, including quantum mechanical effects, the effects of random impurities due to unintentional doping, self-consistent modeling of thermal power dissipation, and the resulting self-heating and degradation of device performance. The development of simulation tools that incorporate lattice heating self-consistently in the loop is of immediate need for the semiconductor industry, to assist in better and faster design of nanoscale devices. One such simulation approach is the topic of this book.

The initial ideas for working on the topic of self-heating came about from a discussion in 2003 between Professors Vasileska, Ferry, and Goodnick at Arizona State University. Shortly after these initial ideas were developed, Professor Katerina Raleva began work on her doctorate with Professors Vasileska and Goodnick, and most of the work presented in chapter 2 is to a large extent derived from the PhD work of Professor Raleva. Soon it was recognized that in nanoscale devices, there is a two-fold need for solving directly the Boltzmann transport equation for phonons in addition to the need for solving the Boltzmann transport equation for carriers (electrons or holes). The first is the requirement to properly account for phonon boundary scattering, and the second one comes from the observation that in highly scaled devices, phonons primarily travel ballistically (in the same manner the carriers do). Hence, a phonon Monte Carlo solver was developed by PhD student Abdul Rawoof Shaik, which is described in chapter 3 of this book.

The book contains four chapters. In chapter 1, we outline some general aspects related to the treatment of self-heating effects in devices (sections 1.1 and 1.2). Modeling of self-heating effects, using the commercial Silvaco ATLAS tool, is briefly described in section 1.3. In chapter 2 we discuss the current state of the art in modeling heating effects in nanoscale devices. We begin, in section 2.1, with some general considerations related to the solution of the heat transport problem in devices. Then, in section 2.2, we describe in detail the ASU thermal particle-based device simulator and present simulation results that are obtained with it. Global modeling that utilizes both SILVACO Atlas and our electro-thermal particle based device simulator to account for a circuit plus interconnects is described in section 2.3. Since in the smallest devices of interest, it is required to solve the Boltzmann transport equation for phonons, the description of the solution is given in chapter 3— section 3.1 (description of phonon–phonon scattering) and section 3.2 (phonon Monte Carlo description). Verification of the phonon Monte Carlo is presented in section 3.3 and simulation results for the thermal conductivity of Si at various temperatures are presented in section 3.4. Conclusive comments related to this work and future directions of research are given in chapter 4. The derivation of the energy balance equations for the optical and the acoustic phonons by taking moments of the phonon Boltzmann transport equation is given in appendix A.

Acknowledgments

Dr Vasileska and Dr Goodnick would like to thank Dr Ferry from Arizona State University for the valuable discussions that initiated this project many years ago. Dr Vasileska and Dr Raleva would also like to thank Dr Erik Bury and Dr Ben Kaczer from IMEC (Belgium) for motivating us to perform global multi-scale simulations that involve our electro-thermal device simulator and GIGA3D module of the Silvaco TCAD tool.

Author biographies

Abdul Rawoof Shaik

Abdul Rawoof Shaik received a bachelor's degree in Electrical Engineering from Indian Institute of Technology Patna in 2012. From 2012 to 2014 he worked in the analog IC design industry, designing transceivers for USB, PCI and SATA protocols. He completed his master's degree in Electrical, Computer and Energy Engineering from Arizona State University in 2016. He worked on modeling thermal transport in nanoscale devices for his master's thesis. He is currently pursuing a PhD in Arizona State University working on modeling defect migration in cadmium telluride solar cells. During his bachelor's studies, he published two conference papers on embedded systems. He received the best project award for his project work on implementing text to speech on ARM based embedded systems. He co-authored four conference papers and two book chapters about thermal modeling of nanoscale devices during his master's thesis work. He received third best poster award at the 2017 NREL/SNL/BNL PV Reliability Workshop.

Dragica Vasileska

Dragica Vasileska received the BSEE (Diploma, equivalent to MS Degree in USA) and the MSEE Degree from the University Sts Cyril and Methodius (Skopje, Republic of Macedonia) in 1985 and 1992, respectively, and a PhD Degree from Arizona State University in 1995. From 1995 until 1997 she held a Faculty Research Associate position within the Center of Solid State Electronics Research at Arizona State University. In the fall of 1997 she joined the Faculty of Electrical Engineering at Arizona State University. In 2002 she was promoted to Associate Professor and in 2007 to Full Professor. Her research interests include semiconductor device physics and semiconductor device modeling, with a strong emphasis on quantum transport and Monte Carlo particle-based device simulations. Recently, her focus has shifted towards modeling metastability and reliability of solar cells. She is a Senior Member of both IEEE and APS. Professor Vasileska has published more than 180 publications in prestigious scientific journals, over 200 conference proceedings refereed papers, dozens of book chapters, has given numerous invited talks and is a co-author on two books: *Computational Electronics*, D Vasileska and S M Goodnick, Morgan & Claypool, 2006; *Computational Electronics: Semiclassical and Quantum Transport Modeling*, D Vasileska, S M Goodnick and G Klimeck, CRC Press, 2010. She is also an editor of two books: *Cutting Edge Nanotechnology*, In-Tech, 2010 and *Nano-Electronic Devices: Semiclassical and Quantum Transport Modeling* (Co-Editor S M Goodnick)

Springer, July 2011. She has many awards including the best student award from the School of Electrical Engineering in Skopje since its existence (1985, 1990). She is also a recipient of the 1998 NSF CAREER Award. Her students have won the best paper and the best poster award at the LDSD conference in 2004, the best oral presentation award at the ISDRS Conference in 2011, two best poster awards at the 2014 IMAPS Symposium, two best poster awards at the 42nd PVSC Conference in 2015, best student oral presentation at the 2015 IMAPS Symposium, best poster at the 2016 IMAPS Symposium, 3rd best poster place at the 2017 NREL/SNL/BNL PV Reliability Workshop, etc.

Katerina Raleva

Katerina Raleva received the MSEE Degree in Electrical Engineering and the PhD degree from the University Sts Cyril and Methodius (Skopje, Republic of Macedonia) in 2002 and 2008, respectively. Her PhD research work (modeling thermal effects in deep sub-micrometer SOI devices) is done with collaboration of the Center of Solid State Electronics Research at Arizona State University. She is currently an Associate Professor within the Institut of Electronics at Faculty of Electrical Engineering and Information Technologies (FEIT), Skopje, Macedonia. Her research interests include, semiconductor physics, semiconductor device modeling and modeling devices on a circuit level. Professor Raleva has published more than 80 scientific publications in scientific journals and conference proceedings and several book chapters.

Stephen M Goodnick

Stephen M Goodnick received his PhD degrees in electrical engineering from Colorado State University, Fort Collins, in 1983. He was an Alexander von Humboldt Fellow with the Technical University of Munich, Munich, Germany, and the University of Modena, Modena, Italy, in 1985 and 1986, respectively. He served as Chair and Professor of Electrical Engineering with Arizona State University, Tempe, from 1996 to 2005. He served as Associate Vice President for Research for Arizona State University from 2006–2008, and presently serves as Deputy Director of ASU Lightworks, and is Hans Fischer Senior Fellow with the Institute for Advanced Studies at the Technical University of Munich. Professionally, he served as President (2012–2013) of the IEEE Nanotechnology Council, and served as President of IEEE Eta Kappa Nu Electrical and Computer Engineering Honor Society Board of Governors, 2011–2012. Some of his main research contributions include analysis of surface roughness at the Si/SiO_2 interface, Monte Carlo simulation of ultrafast carrier relaxation in quantum confined systems, global modeling of high frequency and energy conversion devices, full-band simulation of semiconductor devices, transport in nanostructures, and fabrication and characterization of nanoscale semiconductor devices. He has published over 400 journal articles, books, book chapters, and conference proceeding, and is a Fellow of IEEE (2004) for contributions to carrier transport fundamentals and semiconductor devices.

Chapter 1

Introduction

Richard Feynman's 1959 lecture *There is plenty of room at the bottom* has often been quoted when people talk about nanoscience and nanotechnology [1]. He predicted that 'we will get an enormously greater range of properties that substances can have, and of different things that we can do' if atoms and molecules can be arranged in the way we want. However, the real take-off of nano-related research and technological exploitation started at about 15 years ago [2–5]. This is a logical consequence of the developments of science and technology.

The 20th century has been called the 'century of physics' because of the revolutionary development of physics and its tremendous impacts on society. A solid foundation has been laid down to describe nature at the elementary particle level at one end to the evolution of the Universe at the other. Of close relevance to our life (and economy), quantum mechanics has helped us to reveal the nature of atoms, molecules and solids. Solid state physics led to the creation and great success of semiconductor science and engineering. Integrated circuits, laser and magnetic disks are indispensable to information technology (IT) and our daily life.

Our understanding and exploitation of the material world around us have been pushing forward in two opposite directions: *from the bottom up* [6, 7] and *from the top down*. In the bottom-up approach, we start with electrons and nuclei as the basic building blocks. The properties of atoms and most of the relatively simple molecules (this can be called the sub-nm world) are well understood. At increasing levels of complexity, we deal with macro-molecules, polymers, clusters and bio-molecules (these are relatively small nanostructures which we deal with). On the other hand, *from the top-down approach*, advances in micro-fabrication processes have led to a continuous miniaturization of field effect transistors (FETs) that contain semi-conductors (e.g. silicon), insulators (e.g. silicon dioxide), and metallic (e.g. copper interconnects) layers only a few nanometers thick. For example, in recent years, technology has advanced to fabricate integrated circuits (ICs) at 14 nm gate length commercially [8]. Fabrication industry giants like Intel, TSMC, Samsung and

doi:10.1088/978-1-6817-4123-9ch1 1-1

Global Foundries have plans to fabricate ICs at 7 nm by 2017. Samsung has already fabricated and tested a 128 Mb SRAM with 10 nm feature size [9]. This aggressive scaling of technology is possible because of the advent of FinFETs, fully-depleted silicon-on-insulator (FDSOI) device technology, and other innovations in the microelectronics industry.

Device miniaturization is not without problems, however. More devices per unit area results in increasingly larger amounts of heat generated per unit volume [10]. *Self-heating* may lead to a substantial increase in the effective operating temperature of the device, which degrades the device electrical performance and also affects device reliability. A recent study on self-heating of 14 nm down to 7 nm silicon FinFETs shows that heat confinement in the Si channel increases by 20% and in the strained Ge channel by 57% [11]. This, in turn, results in 70 K and 100 K changes in the channel temperature for 14 nm and 7 nm FETs, respectively. Hence efficient heat removal methods are necessary to increase device performance and device reliability. A recent trend in peak efficiency versus power density of the switched capacitor power converters shows that the efficiency decreases as power density increases [12]. This implies that the efficiency of utilizing the electrical energy in logic operations is greatly reduced as the power density increases.

Self-heating effects are particularly important for transistors in silicon-on-insulator (SOI) technology, where the device is separated from the substrate by a low thermal conductivity buried silicon dioxide layer (figure 1.1), as well as copper interconnects that are surrounded by low thermal conductivity dielectric materials [13]. This, in turn, leads to a substantial elevation of the local device temperature which modifies the device output characteristics. Another important aspect that needs to be considered in nanoscale SOI devices, is a reduction of the thermal conductivity in semiconductor thin films due to surface scattering of phonons. For instance, bulk silicon (Si) has a thermal conductivity of 148 W m^{-1} K^{-1}, while a 10 nm Si-film has around 10 times smaller thermal conductivity value, which is due to the reduction in mean free path of phonons in confined structures.

Since the local lattice temperature is very difficult to measure, being an internal variable, a combination of nanoscale experimental techniques combined with accurate modeling methods must be employed in order to determine the temperature profile in the device, particularly the so-called 'hotspot' where the peak temperature occurs, as shown in figure 1.1. Accurate thermal modeling and the design of

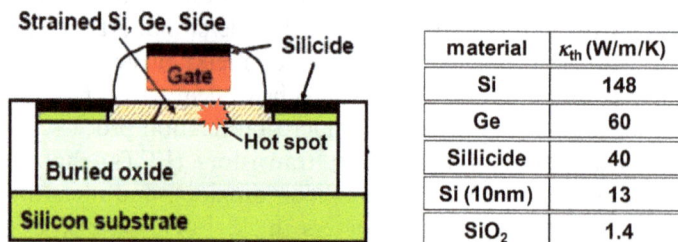

material	κ_{th} (W/m/K)
Si	148
Ge	60
Sillicide	40
Si (10nm)	13
SiO$_2$	1.4

Figure 1.1. Position of the thermal hotspot (left) and a table of the thermal conductivity (right) for materials used in nanoscale SOI devices. Note the decrease of the thermal conductivity of 10 nm Si-film compared to the case of bulk Si.

microelectronic devices and thin film structures at the micro- and nanoscales poses a challenge to electrical engineers who are less familiar with the basic concepts and ideas in sub-continuum heat transport. This book aims to bridge that gap.

1.1 Some general aspects of heat conduction

A medium through which heat is conducted may involve the conversion of mechanical, electrical, nuclear, or chemical energy into heat (or thermal energy). In heat conduction analysis, such conversion processes are characterized as heat (or thermal energy) generation. For example, the temperature of a resistance wire rises rapidly when electric current passes through it as a result of the electrical energy being converted to heat at a rate of I^2R, where I is the current and R is the electrical resistance of the wire. The safe and effective removal of this heat away from the places of heat generation (the electronic circuits) is the subject of electronics cooling, which is one of the modern application areas of heat transfer.

The rate of heat generation in a medium may vary with time as well as position within the medium. We denote with $H(x,t)$ the supplied heating power density for a case of a 1D system in which we consider an element with thickness Δx, density ρ, specific heat of the material c and area A that is normal to the direction of heat transfer. An energy balance on this element during a time interval Δt can be expressed as

$$\begin{pmatrix} \text{Rate of heat} \\ \text{conduction} \\ \text{at } x \end{pmatrix} - \begin{pmatrix} \text{Rate of heat} \\ \text{conduction} \\ \text{at } x + \Delta x \end{pmatrix} + \begin{pmatrix} \text{Rate of heat} \\ \text{generation} \\ \text{inside the} \\ \text{element} \end{pmatrix} = \begin{pmatrix} \text{Rate of change} \\ \text{of the energy} \\ \text{content of the} \\ \text{element} \end{pmatrix}$$

or in mathematical terms:

$$F_x - F_{x+\Delta x} + H(x, t)A\Delta x = \frac{\Delta E_{\text{element}}}{\Delta t}, \tag{1.1}$$

where, F is the rate of heat conduction, that is given by the Fourier law of heat conduction, which in 1D reads:

$$F = -\kappa A \frac{\partial T}{\partial x} \tag{1.2}$$

where κ is the thermal conductivity. The change in energy content of the element is calculated using:

$$\Delta E_{\text{element}} = \rho c A \Delta x (T_{t+\Delta t} - T_t). \tag{1.3}$$

Substituting the result from equation (1.3) into equation (1.1), and dividing by $A\Delta x$ leads to:

$$-\frac{1}{A}\frac{F_x - F_{x+\Delta x}}{\Delta x} + H(x, t) = \rho c \frac{T_{t+\Delta t} - T_t}{\Delta t}. \tag{1.4}$$

Finally, substituting equation (1.2) into equation (1.4), and letting Δt and Δx go to zero, gives the 1D version of the heat conduction equation:

$$\rho c \frac{\partial T}{\partial t} = \frac{\partial}{\partial x}\left(\kappa \frac{\partial T}{\partial x}\right) + H(x, t). \tag{1.5}$$

For the case of heat propagation in a 3D isotropic and inhomogeneous medium, the heat transport equation is of the form:

$$\rho c(\mathbf{r}, T)\frac{\partial T}{\partial t} = C(\mathbf{r}, T)\frac{\partial T}{\partial t} = \nabla \cdot (\kappa(\mathbf{r}, T)\nabla T) + H(\mathbf{r}, t) \tag{1.6}$$

where $C(\mathbf{r},T)$ is position and temperature dependent heat capacity and $H(\mathbf{r},t)$ is the supplied heating power.

As we clearly see from the derivation presented above, the heat conduction equation is a consequence of Fourier's law of cooling. If the medium is not the whole space, in order to solve the heat conduction equation uniquely we also need to specify boundary conditions for T. To determine uniqueness of solutions in the whole space it is necessary to assume an exponential bound on the growth of solutions, this assumption is consistent with observed experiments. Solutions of the heat equation are characterized by a gradual smoothing of the initial temperature distribution by the flow of heat from warmer to colder areas of an object. Generally, many different states and starting conditions will tend toward the same stable equilibrium. As a consequence, to reverse the solution and conclude something about earlier times or initial conditions from the present heat distribution is very inaccurate except over the shortest of time periods. (The heat equation is the prototypical example of a parabolic partial differential equation.)

At nanometer length scales, the familiar continuum Fourier law for heat conduction fails due to both classical and quantum size effects [14–16]. The past two decades have seen increasing attention to thermal conductivity and heat conduction in nanostructures. Experimental methods for characterizing the thermal conductivity of thin films and nanowires have been developed and are still evolving. Experimental data have been reported on various nanostructures: thin films, superlattices, nanowires, and nanotubes. Along the way, models and simulations have been developed to explain the experimental data.

The rest of this section summarizes some past work and the current understanding of heat conduction in nanostructures. We first give a brief overview of the fundamental physics that distinguishes phonon heat conduction in nanostructures from that in macrostructures. Then we discuss some of the size effects in nano-structures that impact their thermal conductivity.

Heat conduction in dielectric materials and most semiconductors is dominated by lattice vibrational waves. The basic energy quantum of lattice vibrations is called a phonon, analogous to a photon which is the basic energy quantum of an electromagnetic wave. Similar to photons, phonons can be treated as both waves and particles. Size effects appear if the structure characteristic length is comparable to or smaller than the phonon characteristic lengths. Two kinds of size effects can exist: the classical size effect, when phonons can be treated as particles, and the wave effect, when the wave phase information of phonons becomes important. Distinction between these two regimes depends on several characteristic lengths [17]. In this context, the important characteristic lengths of phonon heat conduction are the mean free path, the wavelength, and the phase coherence length [18]. The mean free path is the average distance that phonons travel between successive collisions. The mean free path Λ is often estimated from kinetic theory and is used in the calculation of the thermal conductivity of the material/structure of interest as:

$$\kappa = \frac{1}{3}C_V v_P \Lambda \qquad (1.7)$$

In equation (1.7) C_V and v_P are the volumetric specific heat capacity of a phonon, and the phonon velocity inherent in a material. In silicon, for example, the phonon mean free path is on the order of ~300 nm [19] at room temperature in bulk materials. The phase of a wave can be destroyed during collisions, which is typically the case in inelastic scattering processes, such as phonon–phonon scattering.

Figure 1.2 compares the dimensions of several nanostructures (e.g. an SOI device and a superlattice structure) with the dominant phonon mean free path (MFP) and wavelength at room temperature. The graph also provides a general guideline for the appropriate treatment of phonon transport in nanostructures. Phonon transport can be predicted using the Boltzmann transport equation (BTE) for each type of phonon, which are required only when the scattering rates of electrons or phonons vary significantly within a distance comparable to their respective mean free paths.

The phonon BTE simply takes care of the bin counting of the energy of carrier particles of a given velocity and momentum, scattering in and out of a control

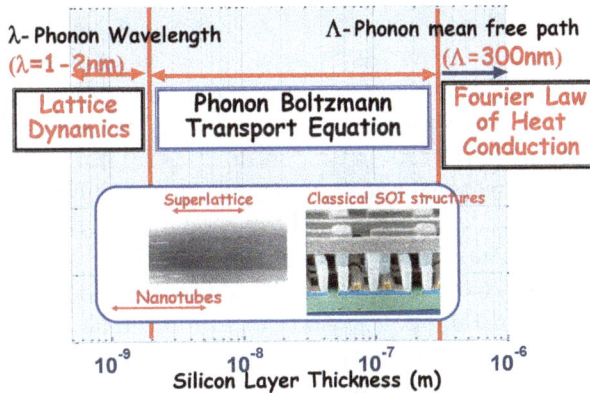

Figure 1.2. Regime map for phonon transport in ultra-thin silicon layers. The mean free path, Λ, is the distance that phonons travel on average before being scattered by other phonons or crystal defects. If the dimensions of the silicon layer are smaller than Λ, the phonon Boltzmann transport equation should be used for heat transfer analysis of the thin film. The dominant phonon wavelength, λ, at room temperature, is on the order of 2–3 nm. Analogously, phonon wave simulations should be performed for devices with thicknesses comparable to λ.

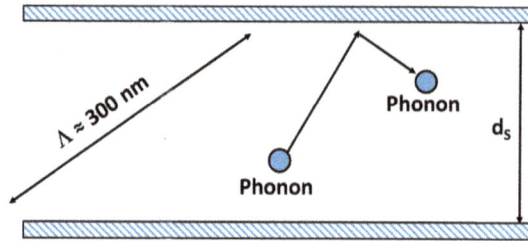

Figure 1.3. Phonon-boundary scattering is responsible for a large reduction in the thermal conductivity of a thin silicon layer where the thickness of the film, d_s, is comparable to or smaller than the phonon mean free path, Λ. $T = 300$ K is assumed in this example.

volume at a point in space and time. Analysis of the heat transfer in microelectronic devices, interconnects and nanostructures using the BTE is very cumbersome and complicated, even for simple geometries, and has been the topic of research and development in the field of micro- and nanoscale heat transfer for the past two decades (e.g. [20]). Equation (1.7) can only be considered as providing the qualitative behavior of a thermal conductivity from which the thermal conductivity is found to be proportional to the phonon mean free path. The phonon MFP is well known to become shorter when the system gets hotter because the phonon population is increased, which causes the collision frequency among phonons to be higher. Increased phonon collisions prevent the phonons with high energy in the hot region from moving to the cold region and vice versa. This means that the energy transport is low; consequently the thermal conductivity is low. Therefore, it can be inferred that phonon scattering governs the thermal conductivity.

Detailed descriptions and analyses of the ballistic heat transfer in a semiconductor/metallic layer are beyond the scope of this book. However, the most prominent manifestation of ballistic heat transport in thin films would occur in the form of large reductions in thermal conductivity compared to the bulk values. Ballistic phonon transport in silicon films, or phonon-boundary scattering (see figure 1.3), has been investigated through large measured reductions in the lateral thermal conductivity compared to the bulk value near room temperature [21–23] and is shown in figures 1.4 and 1.5 for a sheet of a material and for a nanowire, respectively.

The lateral thermal conductivity of the thin silicon layer decreases as the thickness of the film is reduced. Deviation of the thermal conductivity from the bulk value takes a sharp dive as the thickness of the film is reduced beyond 300 nm, which is the order of magnitude for the phonon mean free path in silicon at room temperature. For example, the thermal conductivity of the 20 nm thick silicon layer is nearly an order of magnitude smaller than the bulk value. It should also be mentioned that the Fourier heat conduction equation cannot explain the thickness dependency of thermal conductivity in silicon. The impact of phonon-boundary scattering on the thermal conductivity of a thin silicon layer can be predicted (see figure 1.4), using the BTE and the theory described by Asheghi *et al* [19], such that it agrees very well with the experimental data.

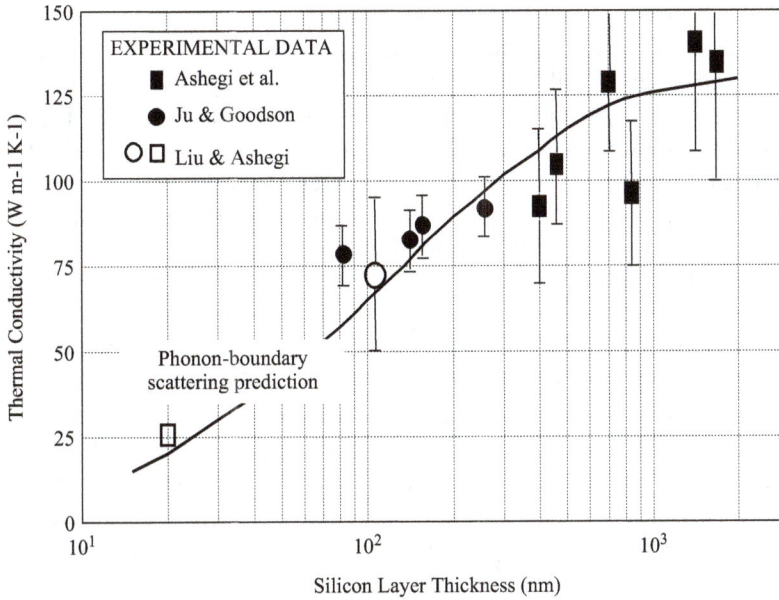

Figure 1.4. Room-temperature thermal conductivity data for silicon film layers as a function of thickness [22].

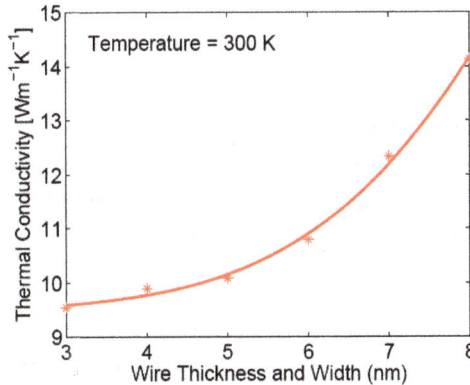

Figure 1.5. Variation of the thermal conductivity of silicon nanowires with cross section. Note the significant drop in the thermal conductivity from its bulk value.

One way to roughly estimate the impact of the micro/nanoscale effect is to use the modified thermal conductivity values for thin silicon and copper layers in conventional thermal simulation tools that use the continuum theory or diffusion equation. However, one should be cautious that these tools can by no means capture the non-equilibrium energy transport in nanostructures and provide only an estimate for the effective thermal resistance of a given device. At the same time, the temperature distribution and heat flux patterns under non-equilibrium conditions could be entirely different from those predicted by the Fourier heat conduction equation. It should also be mentioned that the proposed approach should be implemented with extra care and with a full understanding and mastery of sub-continuum heat transport effects and concepts. The thermal boundary resistance at the interfaces of high thermal conductivity materials, such as copper–silicon (metal–dielectric) and silicon–germanium (dielectric–dielectric), are often the dominant thermal resistance in the heat conduction path in micro- and nanoscale devices and structures, and should be carefully accounted for in the estimation of the total thermal resistance of a given device. For example, the contact resistance between metal–dielectric layers can vary between 5×10^{-9} and 5×10^{-7} m^2 K W^{-1}, depending on the processing conditions and quality of the interface [24], which are comparable to those of 1 and 100 μm thick silicon layers, respectively. The thermal contact resistance between dielectrics, such as Si and Ge in SiGe superlattices (for a thermoelectric application), could be as small as 5×10^{-11} to 5×10^{-10} m^2 K W^{-1} [25]; however, the total thermal resistance of the SiGe superlattice consists of hundreds of SiGe bi-layers and thus could be relatively large. Phonon transport at the interface of single crystalline dielectrics layers is one of the most exciting and heavily studied topics in the field of micro- and nanoscale heat transfer (e.g.) [26, 27].

In the ultra-fast laser heating processes at time scales of 10^{-15} to 10^{-12} seconds, as well as high speed transistors switching at timescales in the order of 10^{-11} seconds, the temperatures of the electron and phonon systems are not in equilibrium and may differ by orders of magnitude. Even after the phonon and electron reach equilibrium, the energy carried away by phonons can travel only to 10–100 nanometers; therefore, the temperature of the transistor can easily rise to several times its designed reliability limit. Under these circumstances, regardless of the cooling solution at the packaging level, a catastrophic failure at the device level can occur, because the impact of the rapid temperature rise is limited to the device and its vicinity. As a result, while the package level cooling solutions can reduce the quasi-steady-state/average temperature across a microprocessor or at length scales of the order of one millimeter, it has very little impact at micro/nanoscales. Basically, there is no practical way to reduce the temperature at the device and interconnect level by means of a cooling device or solution; therefore, the options for thermal engineering of these devices are very limited. However, intelligent electro-thermal design along with careful floor planning at the device level can largely reduce the temperature rise within a device. This means that the role of the thermal engineer is to properly anticipate—perhaps in full collaboration with electrical engineers—and prevent the problem at the early stages and at the device level, rather than to pass the problem to the package-level thermal engineers.

1.2 Solution of the self-heating problem

To properly treat heating without any approximations made in the problem at hand, one in principle has to solve the coupled Boltzmann transport equations for the electron and phonon systems together. More precisely, *one has to solve the coupled electron—optical phonons—acoustic phonons—heat bath problem*, where each sub-process involves different time scales and has to be addressed in a somewhat individual manner and included in the global picture via a self-consistent loop. We consider the coupled system of semi-classical Boltzmann transport equations for the distribution functions of electrons $f(\mathbf{r},\mathbf{k},t)$ and phonons $g(\mathbf{r},\mathbf{k},t)$:

$$\left(\frac{\partial}{\partial t} + v_e(\mathbf{k}) \cdot \nabla_r + \frac{e}{\hbar}E(\mathbf{r}) \cdot \nabla_{\mathbf{k}}\right)f$$
$$= \sum_q \left\{ W_{e,q}^{k+q\rightarrow k} + W_{a,-q}^{k+q\rightarrow k} - W_{e,-q}^{k\rightarrow k+q} - W_{a,q}^{k\rightarrow k+q} \right\} \qquad (1.8a)$$

$$\left(\frac{\partial}{\partial t} + v_p(q) \cdot \nabla_r\right)g = \sum_k \left\{ W_{e,q}^{k+q\rightarrow k} - W_{a,q}^{k\rightarrow k+q} \right\} + \left(\frac{\partial g}{\partial t}\right)_{p-p} \qquad (1.8b)$$

Here $W_{e,q}^{k+q\rightarrow k}$ is the probability for electron transition from $\mathbf{k}+\mathbf{q}$ to $\mathbf{k}$ due to emission of phonon q. Similarly $W_{a,q}^{k+q\rightarrow k}$ refers to processes of absorption. The system is nonlinear, as the probabilities W depend on the product $f \bullet g$ of the electron and phonon distribution functions. The last term on the right-hand side of equation (1.8b) accounts for the phonon–phonon interaction. During the evolution of the system, the electrons gain energy from the electrical field E in the device. The transfer of energy between electrons and phonons is due to the terms W, with a timescale of the order of 0.1 ps (see figure 1.6 for more details). These terms are common for equations (1.8a) and (1.8b). Nevertheless, even without the phonon–phonon interaction in (1.8b), the equation set poses a multi-scale problem since the left-hand sides involve different time scales: the velocity v_p of the phonons is two orders of magnitude lower than the velocity v_e of the electrons. Accordingly, the heat transfer by the lattice is a much slower process than the charge transfer.

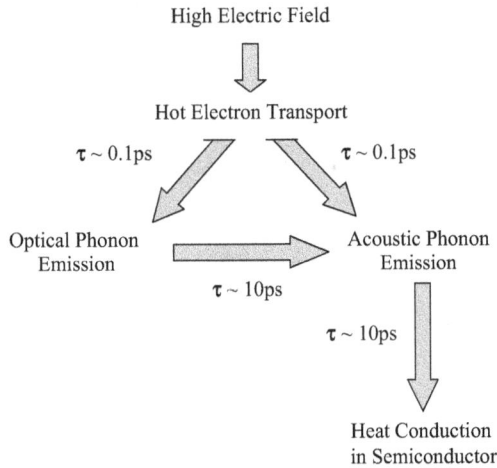

Figure 1.6. The most likely path between energy carrying particles in a semiconductor device is shown together with the corresponding scattering time constants.

Figure 1.6 shows the primary path of thermal energy transport and the associated time constants. With the application of a voltage, electrons gain energy from the electric field and interact with both the optical phonon and the acoustic phonon bath. Note that when considering the electron lattice coupling, the energy transfer from electrons to the high-energy optical phonons is very efficient. However, optical phonons possess negligible group velocity and, thus, do not participate significantly in the heat diffusion. They instead must transfer their energy via anharmonic decay processes to acoustic phonons, which diffuse heat. The energy transfer between phonons is relatively slow compared to electron-optical phonon transport and, thus, thermal non-equilibrium may also exist between optical and acoustic phonons. Therefore, a hot-spot forms.

The general solution of the phonon BTE is rather difficult as it involves multi-particle interactions (such as three-phonon processes, four-phonon processes, etc [28]). Therefore, approximations are needed, and one choice that has been exploited in the literature is the relaxation time approximation for the collision integral on the RHS of the phonon BTE [29], as described in section 3.2 of chapter 3 of this book.

One alternative is to calculate moments of the phonon BTE (section 3.1) and obtain so-called energy-balance equations [30]. Further simplifications of the problem lead to a method that exploits the equations of phonon energy transfer (EPRT) [31], ballistic-diffusive equations (BDE) [32], non-local formulation [33], etc. The other choice is classical molecular dynamics [34] which involves statistical mechanics to compute transport coefficients. Molecular dynamics does not assume the validity of the BTE, but it can be used to compute physical parameters needed by the BTE. Figure 1.7 summarizes the hierarchy of semi-classical thermal phonon modeling. We focus on the methods that have been implemented at Arizona State University in the rest of this section and, thus, highlight the main features of the energy-balance (chapter 2) and the MC solution of the phonon BTE in the relaxation time approximation (chapter 3).

According to figure 1.6, the primary path of energy transport is represented first by scattering between electrons and optical phonons (T_{LO}) and then optical phonons to the lattice (T_A) [35]. The BTE for the two kinds of phonons can then be used to provide the energy balance of the process. This means that one can couple the electron BTE with the equations for the optical and the acoustic energy transfer (that are derived from the phonon BTE in section 3.3 using energy balance and in appendix A taking moments of the phonon BTE) of the form [28]:

$$C_{LO}\frac{\partial T_{LO}}{\partial t} = \frac{3nk_B}{2}\left(\frac{T_e - T_L}{\tau_{e-LO}}\right) + \frac{nm^*v_d^2}{2\tau_{e-LO}} - C_{LO}\left(\frac{T_{LO} - T_A}{\tau_{LO-A}}\right), \qquad (1.9a)$$

$$C_A\frac{\partial T_A}{\partial t} = \nabla \cdot (k_A\nabla T_A) + C_{LO}\left(\frac{T_{LO} - T_A}{\tau_{LO-A}}\right) + \frac{3nk_B}{2}\left(\frac{T_e - T_L}{\tau_{e-L}}\right). \qquad (1.9b)$$

The first two terms in the right-hand side (RHS) of (1.9a) represent the energy gain from the electrons, where n is the electron density and v_d is the drift velocity, while

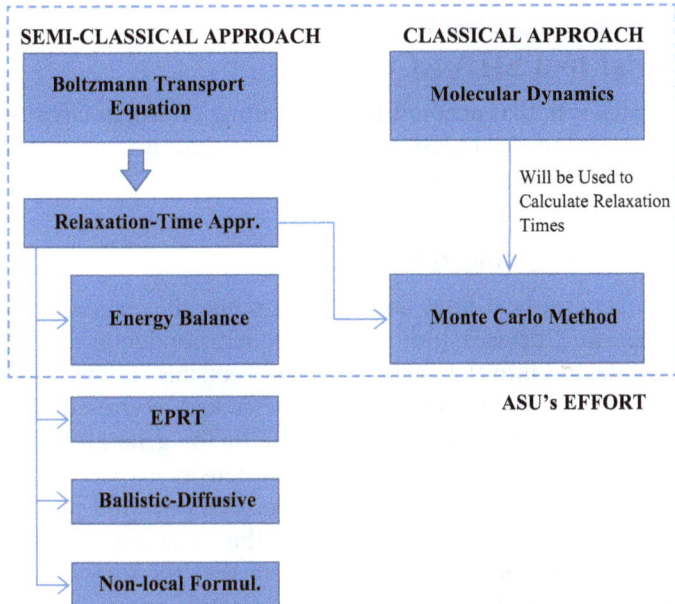

Figure 1.7. Solutions to the phonon transport problem in the literature and at Arizona State University (ASU). They can be categorized into semi-classical (direct or indirect solution of the phonon BTE) and classical approaches based on real-space molecular dynamics calculations. The ASU approach thus far has been to use the relaxation time approximation in conjunction with the phonon BTE and derive energy balance equations for the acoustic and optical phonon baths. These are then solved self-consistently with the Monte Carlo–Poisson solver for the electron system.

the last term is the energy loss to the acoustic phonons. The latter appears as a gain term on the RHS of (1.9b). The first term on the RHS of (1.9b) accounts for the heat diffusion and the last term must be excluded if the electron–acoustic phonon interaction is treated as elastic. In this term, the lattice temperature, T_L, can be estimated as equivalent to T_A. Note that proper boundary conditions accounting for the heat sink apply for (1.9b). C_{LO} and C_A represent the heat capacity of optical and acoustic phonons, respectively, and k_A is the thermal conductivity.

To reiterate, we have used a Chapman–Enskog type expansion [36] to replace the microscopic phonon transport equation by a diffusion problem for the local density and energy of the phonons, where the diffusion coefficients are dependent on the state of the electrons. This method involves computation of the phonon energy dependent scattering tables in the ensemble Monte Carlo (EMC) code for the electron transport, which already represents a big improvement over the current state of the art, where, as already discussed, the coupling of thermal and charge effects is strictly one way. This approach also takes care of the multi-scale nature of the problem, assuming a quasi-steady-state: the phonons are in steady state, albeit with a spatially dependent temperature distribution.

1.3 Modeling heating effects in state of the art devices with the commercial tool SILVACO

There are two packages that account for self-heating effects in circuits (structures)/ devices: Thermal3D and GIGA/GIGA3D package that are an integral part of the SILVACO Atlas device simulation software.

1.3.1 Thermal3D package from Silvaco

Thermal3D is a general heat-flow simulation module that predicts heat-flow from any power generating devices (not limited to semiconductor devices), typically through a substrate and into the package and/or heatsink via the bonding medium. Operating temperatures for packaged and heat sinked devices or systems can be predicted for the design and optimization phase or for general system analysis. It solves equation (1.6), where *H*, the heating power density is spatially uniform within each region and may also have a time-evolution. The time and position dependent temperature distribution is obtained by solving this equation with some boundary conditions. The specific boundary conditions are that the temperature is fixed on some electrodes. For boundaries where the temperature is not fixed, the heat flow

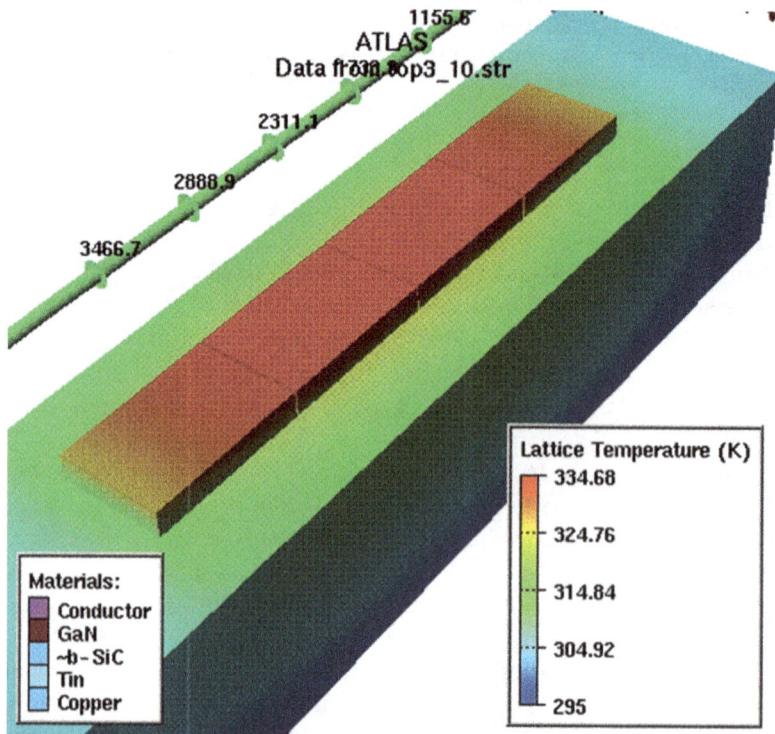

Figure 1.8. Temperature simulation in Thermal3D of GaN HEMT devices this time fabricated onto a β-silicon carbide substrate mounted onto a copper heat sink. (www.silvaco.com). Copyright@Silvaco.

across the boundary is assumed to be zero. In the case when the heat-generation term H is independent of time, the heat transport equation simplifies to the Poisson-like equation:

$$\nabla \cdot (\kappa(\mathbf{r}, T)\nabla T) = -H(\mathbf{r}) \qquad (1.10)$$

whose solution gives the steady-state temperature distribution. Thus, the key features of the Thermal3D simulator are:

- Predicts heat flow and temperature rise for material systems and any number of heat generating sources.
- Models are validated using measured data.
- Three models for heat dependent thermal conductivity to choose from for each of the materials in the system.
- User definable thermal conductivities and coefficients for each material.
- Very fast simulation times allow many combinations to be tried for system design optimization.

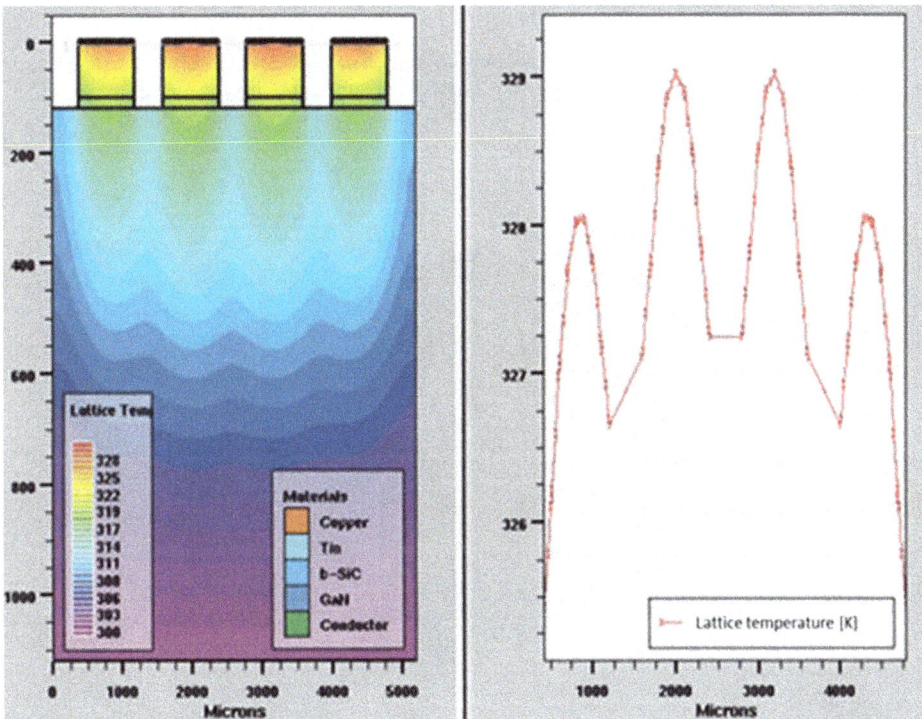

Figure 1.9. In this example, Thermal3D was used to investigate the optimal spacing between the devices and the effect on operating temperature profile. Here, a cut line through the center of the devices is shown together with surface temperature. (www.silvaco.com). Copyright@Silvaco.

- Seamlessly integrated into Silvaco's device simulation software framework, ATLAS.
- Industry leading, easy to use, multi-dimensional visualization tools for results analysis.
- Interactive, user friendly and flexible runtime environment for quick result generation and analysis with numerous examples.

Some examples that demonstrate the usefulness of the Thermal 3D simulation package are given in figures 1.8 and 1.9, respectively.

1.3.2 Giga—non-isothermal device simulator

The Giga/Giga3D module (referred to as Giga) incorporates the effects of self-heating into device simulation. Giga implements Wachutka's thermodynamically rigorous model of lattice heating [37], which accounts for Joule heating, heating, and cooling due to carrier generation and recombination, and the Peltier and Thomson effects. Giga accounts for the dependence of material and transport parameters on the lattice temperature. Giga also supports the specification of general thermal

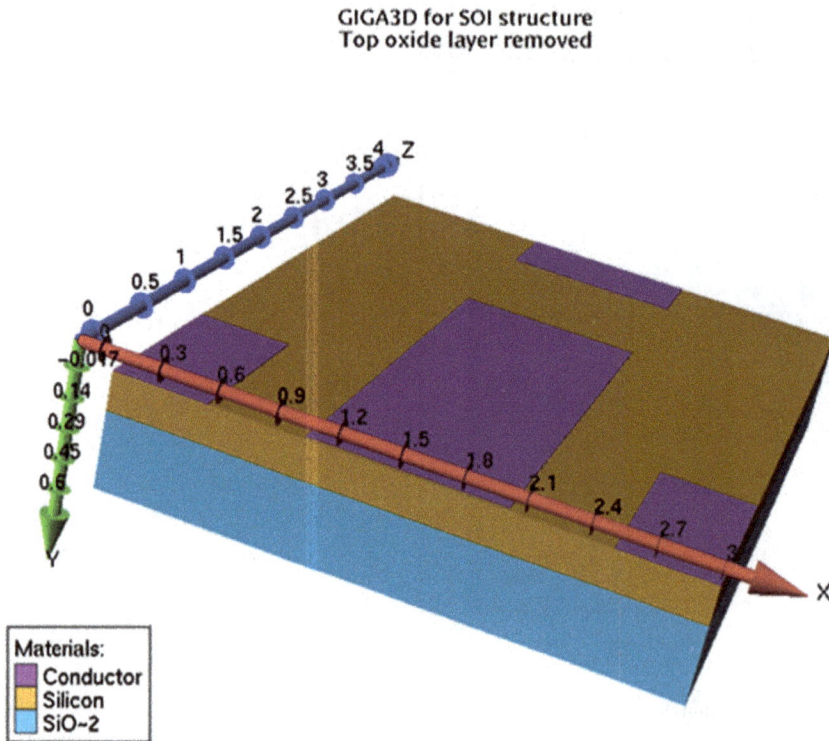

Figure 1.10. Device structure of a short channel ultra-thin SOI transistor with a body contact. The top oxide layer has been removed for clarity. Silicon thickness is 0.2 microns and effective channel length is 0.8 microns. (www.silvaco.com). Copyright@Silvaco.

environments using a combination of realistic heat-sink structures, thermal impedances, and specified ambient temperatures. Physical and model parameters become dependent on the local lattice temperature where appropriate, allowing the self-consistent coupling between the semiconductor device equations and the lattice temperature. Key features of the Giga simulator are:

- Self-consistent lattice temperature solver.
- Coupled to drift-diffusion or hydrodynamic equations.
- Able to model steady state, transient and small signal a.c. biasing.
- Default parameters for thermal conductivity over a range of materials.
- Default parameters for heat capacity over a range of materials.
- Lattice temperature dependence for a wide range of parameters.
- Joule and Peltier/Thomson heat generation terms.
- Flexible boundary condition specification.
- Choice of non-linear solvers for coupling to drift-diffusion equations.
- Anisotropic thermal conductivity tensor.
- Flexible thermopower specification, including phonon drag.

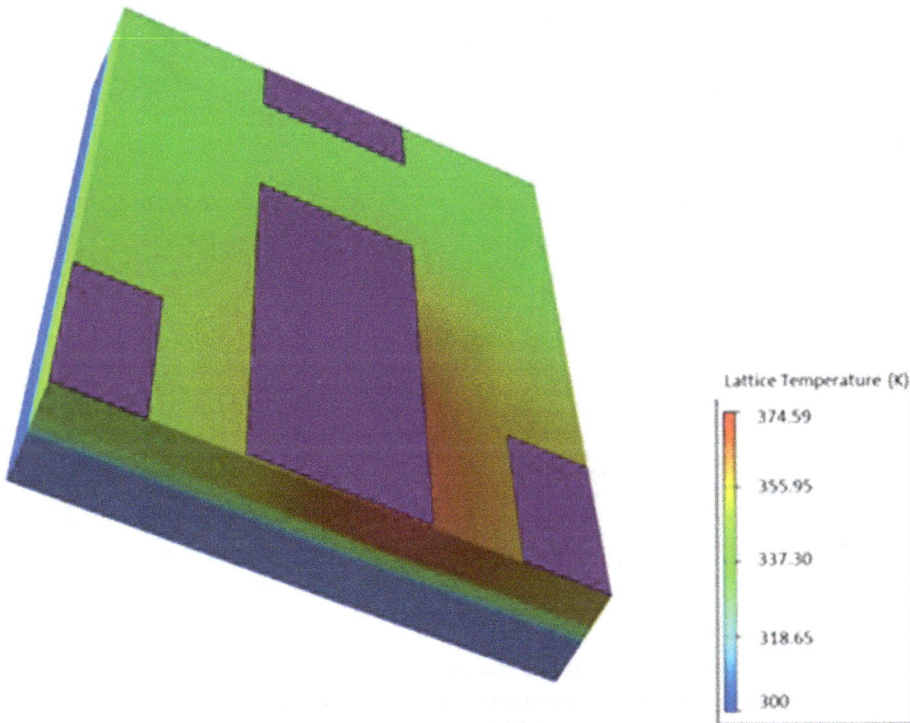

**GIGA3D for SOI Structure:
3V gate bias; 4V drain bias**

Figure 1.11. Lattice temperature distribution for the SOI transistor at a gate bias of 3 V and a drain bias of 4 V. (www.silvaco.com). Copyright@Silvaco.

Possible uses of the solver include:

- Simulation of power devices such as rectifiers, thyristors, mosfets, bipolar transistors.
- Modeling LED, SOI, HBT, HEMT devices.
- Thermo-voltaic device simulation.
- Thermal runaway modeling.
- Device efficiency modeling.

Giga takes account of all forms of heat generation within the device. Joule heating, generation-recombination and Peltier Thomson heat effects are self-consistently solved with all the semiconductor equations. Giga can be used for all DC, AC and transient simulations. In Giga's simplest applications, such as for power devices, Joule heating is often all that is needed, as shown in the following examples. In the Silvaco Giga module, the heat conduction equation is coupled to the Joule heating term with either the drift-diffusion or energy balance equations of the carriers. This then leads to the so-called non-isothermal drift-diffusion or energy balance models [38–40]. The coupling between the electron/hole transport in devices

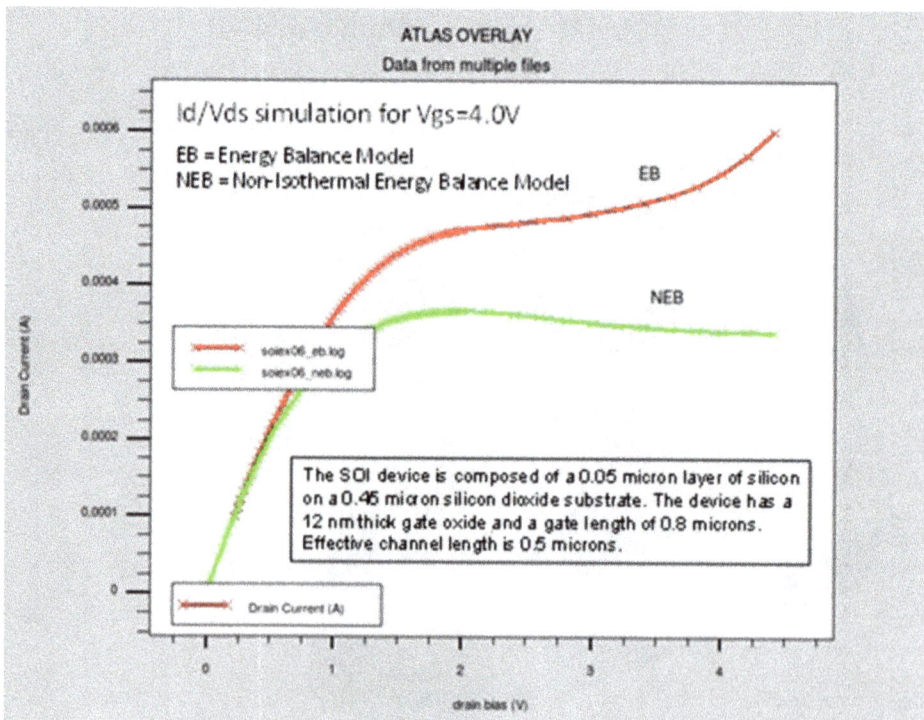

Figure 1.12. Typical characteristics for the SOI transistor are shown, both with and without the Giga3D lattice temperature model enabled. With the increase of drain bias, lattice temperature increases leading to a reduction of the mobility and thus a reduction of the current. This phenomenon is called negative differential resistance (NDR) and can only be correctly simulated in SOI devices with a lattice heating model. (www.silvaco.com). Copyright@Silvaco.

and the corresponding heat flow is achieved via temperature dependent mobilities and diffusion coefficients. Thus, on one hand, the lattice temperature enters the expression for the local mobility value which, in turn, affects the electrostatics and the current density in the device. On the other hand, lattice temperature affects the local Joule heating term which, in turn, affects the lattice temperature profile. The electrical conduction and the heat flow equations are then self-consistently solved for the electrostatic potential and the temperature, respectively.

An application of the Giga3D simulator is given by the example of a short channel ultra-thin SOI transistor with a body contact. Representative results obtained with Giga3D for the device structure being simulated, the lattice temperature distribution and the typical characteristics for the SOI transistor with and without inclusion of lattice degradation of the electron mobility, are shown in figures 1.10–1.12.

References

[1] Feynman R P 1961 *Miniaturization* ed H D Gilbert (New York: Reinhold) pp 282–96
[2] For 2003 International Technology Roadmap for Semiconductors (ITRS), see website http://public.itrs.net/
[3] Keyes R W 2001 Fundamental limits of silicon technology *Proc. IEEE* **89** 227
[4] Nalwa Hari Singh (ed) 2002 *Nanostructured Materials and Nanotechnology* (London: Academic)
[5] Dresselhaus M S and Thomas I L 2001 Alternative energy technologies *Nature* **414** 332
[6] Björk M T, Ohlsson B J, Sass T, Persson A I, Thelander C, Magnusson M H, Deppert K, Wallenberg L R, and Samuelson L 2002 One-dimensional steeplechase for electrons realized *Nano Lett.* **2** 87–9
[7] Datta S 2005 *Quantum Transport: Atom to Transistor* (Cambridge: Cambridge University Press)
[8] Fayneh E, Yuffe M, Knoll E, Zelikson M, Abozaed M, Talker Y, Shmuely Z and Rahme S A 2016 14 nm 6th-generation core processor SoC with low power consumption and improved performance *IEEE Int. Solid-State Circuits Conf.* pp 72–4
[9] Song T *et al* 2016 A 10 nm FinFET 128 mb SRAM with assist adjustment system for power, performance, and area optimization *IEEE Int. Solid-State Circuits Conf.* pp 306–9
[10] Mittal A and Mazumder S 2010 Monte Carlo study of phonon heat conduction in silicon thin films including contributions of optical phonons *J. Heat Transfer* **132** 052402
[11] Jang D *et al* 2015 Self-heating on bulk FinFET from 14 nm down to 7 nm node *IEEE Int. Electron Devices Meeting (IEDM)* pp 11.6.1–11.6.4
[12] International solid-state circuits conference trends 2015 http://isscc.org/trends/
[13] Borkar S 1999 Design challenges of technology scaling *IEEE Micro* **19** 23–9
[14] Geppert L 1999 Solid state [semiconductors. 1999 technology analysis and forecast] *Spectrum, IEEE Spectrum* **36** 52–6
[15] Zeng G, Fan X, LaBounty C, Croke E, Zhang Y, Christofferson J, Vashaee D, Shakouri A and Bowers J E 2003 Cooling Power Density of SiGe/Si Superlattice Micro Refrigerators *Materials Research Society Fall Meeting 2003, Proc.* **793** S2.2 (Boston, MA)
[16] Majumdar A 1993 Microscale heat conduction in dielectric thin films *J. Heat Transfer* **115** 7–16
[17] Chen G 2001 Ballistic-diffusive heat-conduction equations *Phys. Rev. Lett.* **86** 2297–300
[18] Reif F 1985 *Fundamentals of Statistical and Thermal Physics* (London: McGraw-Hill)

[19] Asheghi M, Touzelbaev M N, Goodson K E, Leung Y K and Wong S S 1998 Temperature dependent thermal conductivity of single-crystal silicon layers in SOI substrates *ASME J. Heat Transfer* **120** 30–3

[20] Choi S-H and Maruyama S 2003 Evaluation of the phonon mean free path in thin films by using classical molecular dynamics *J. Korean Phys. Soc.* **43** 747–53

[21] Ju Y S and Goodson K E 1999 Phonon scattering in silicon films with thickness of order 100 nm *Appl. Phys. Lett.* **74** 3005–7

[22] Liu W and Asheghi M 2004 Phonon-boundary scattering in ultra-thin single-crystal silicon layers *Appl. Phys. Lett.* **84** 3819–21

[23] Liu W and Asheghi M 2005 Thermal conductivity of ultra-thin single crystal silicon layers *J. Heat Transfer* **128** 75–83

[24] Ruxandra M, Costescu M, Wall A and Cahill D G 2003 Thermal conductance of epitaxial interfaces *Phys. Rev.* B **67** 054302

[25] Chen G and Shakouri A 2002 Heat transfer in nanostructures for solid-state energy conversion *J. Heat Transfer* **124** 242–52

[26] Vashaee D and Shakouri A 2004 Electronic and thermoelectric transport in semiconductor and metallic superlattices *J. Appl. Phys.* **95** 1233–45

[27] Vashaee D and Shakouri A 2010 Nonequilibrium electrons and phonons in thin film thermionic coolers *Microscale Thermophys. Eng.* **8** 91–100

[28] Ziman J 2001 *Electrons and Phonons: The Theory of Transport Phenomena in Solids* (Oxford: Oxford University Press)

[29] Hardy R J 1970 Phonon Boltzmann equation and second sound in solids *Phys. Rev.* B **2** 1193–206

[30] Mazumder S and Majumder A 2001 Monte Carlo study of phonon transport in solid thin films including dispersion and polarization *J. Heat Transfer* **123** 749–59

[31] Joshi A A and Majumdar A 1993 Transient ballistic and diffusive phonon heat transport in thin films *J. Appl. Phys.* **74** 31–9

[32] Chen G 2002 Ballistic-diffusive equations for transient heat conduction from nano to macroscales *J. Heat Transfer* **124** 320–8

[33] Chen C 1996 Non-local and non-equilibrium heat conduction in the vicinity of nanoparticles *ASME J. Heat Transfer* **118** 539–45

[34] Stillinger F H and Weber T A 1985 Computer simulation of local order in condensed phases of silicon *Phys. Rev.* B **31** 5265–71

[35] Tien C L, Majumdar A and Gerner F M (ed) *Microscale Energy Transport* (London: Taylor and Francis)

[36] Cercignani C 1988 *The Boltzmann Equation and its Applications* vol. 67 *Applied Mathematical Sciences* (Berlin: Springer)

[37] Wachutka G K 1990 Rigorous thermodynamic treatment of heat generation and conduction in semiconductor device modeling *IEEE Trans. Comp. Aided Design* **11** 1141–9

[38] Johnson R G, Snowden C M and Pollard R D 1997 A physics-based electro-thermal model for microwave and millimeter-wave HEMTs *IEEE MTT-S Int. Microwave Symp. Dig.* **3** 1485–8

[39] Batty W, Panks A J, Johnson R G and Snowden C M 2000 Electro-thermal modelling of monolithic and hybrid microwave and millimeter-wave ICs *VLSI Design* **10** 355–89

[40] Silvaco ATLAS manual http://www.silvaco.org

Chapter 2

Current state of the art in modeling heating effects in nanoscale devices

2.1 Some general considerations about the solution of the heat transport problem in devices

The solution of the heat transport problem in semiconductor devices has focused on several issues: (1) independent analysis of the phonon heat bath problem via direct solutions of the phonon Boltzmann transport equations (BTEs), and (2) inclusion of lattice heating in device simulators such as Silvaco by solving the Joule heating (Fourier law) equation without consideration of the microscopic nature of the heat flow and the non-equilibrium state that exists between the acoustic and optical phonons. Recently, there have been attempts to include lattice heating in device simulators by coupling the hydrodynamic equations for the electrons and the energy balance equations for the acoustic and optical phonons [1]. Also, Pop and his co-workers coupled lattice heating with the Boltzmann transport equation for the electrons, including the full phonon dispersion in the picture and for a 1D structure (an $n+$-n-$n+$ diode) [2].

Lai and Majumder [3] developed a coupled electro-thermal model for studying thermal non-equilibrium in submicron silicon MOSFETs. Their results showed that the highest electron and lattice temperatures occur under the drain side of the gate electrode, which, also corresponds to the region where non-equilibrium effects such as impact ionization and velocity overshoot are a maximum. Majumdar *et al* [4] have analysed the variation in performance caused by hot electrons and associated hot phonon effects in GaAs MESFETs. These hot electrons and phonons together were observed to decrease the output drain current by as much as 15%. Thus, they concluded that both electron and lattice heating should be included in the electrical behavior of devices.

doi:10.1088/978-1-6817-4123-9ch2

It has already been recognized (equation (1.6)) that the simulation of devices operated under non-isothermal conditions is of growing importance, and a heat flow equation

$$C\frac{\partial T}{\partial t} = \nabla \cdot (\kappa \nabla T(\mathbf{r}, t)) + H(\mathbf{r}, t) \tag{2.1}$$

where C is the total heat capacity, κ is the thermal conductivity and H is the heat generation rate, is added to conventional drift-diffusion and/or hydrodynamic models to account for the mobility degradation due to lattice heating. Also, an excellent discussion on the form of the heat generation term is detailed in the paper by Wachutka [5]. Briefly, three different models are most commonly used and these include: (1) *Joule Heating*, (2) *electron-lattice scattering* and the (3) *phonon model*. Although these three models yield identical results in equilibrium, under non-equilibrium conditions, the results of the three models can vary significantly. Below we outline the major characteristics of each of these three models.

Case 1: Within the Joule heating model, the thermal model consists of the heat diffusion equation using a local Joule heating term as the source. The source term is computed from the electrical solution as the product of the local field and the current density [6]

$$H = \mathbf{J} \cdot \mathbf{E}. \tag{2.2}$$

This source term is similar to the one used by Leung and co-workers [7] and assumes that recombination heating is negligible. In this case the 'hot spot' will occur near the location where the dot product of the field and of the current density is the largest. Simulations that have used this expression as a heating term suggest that the bulk of the heating will occur directly under the gate region where most of the voltage drop occurs, and where the current density is the largest because of the restricted electron flow path due to the depletion regions. A study by Raman and co-workers on lightly doped drain (LDD) devices suggested that the location of the hot-spot occurs at the drain side of the gate. The complete Leung and co-workers expression used for the source term is

$$H = \mathbf{J} \cdot \mathbf{E} + (R - G)(E_G + 3k_B T) \tag{2.3}$$

where the second term represents the heating rate due to non-radiative generation (G) and recombination (R) of electron–hole pairs. E_G is the semiconductor band-gap, k_B is the Boltzmann constant and T is the lattice temperature.

The above model has been generalized by Waschutka [5] and implemented in the GIGA/GIGA3D Silvaco module, as discussed in chapter 1, to include Joule heating, heating and cooling due to carrier generation and recombination, and the Peltier and Thomson effects:

$$H = \frac{|\mathbf{J}_n|^2}{q\mu_n n} + \frac{|\mathbf{J}_p|^2}{q\mu_p p} - T_L(\mathbf{J}_n \cdot \nabla P_n) - T_L(\mathbf{J}_p \cdot \nabla P_p)$$

$$+ q(R - G)\left[T_L\left(\frac{\partial \phi_n}{\partial T_{n,p}}\right) - \phi_n - T_L\left(\frac{\partial \phi_p}{\partial T_{n,p}}\right) + \phi_p \right] \qquad (2.4)$$

$$- T_L\left[\left(\frac{\partial \phi_n}{\partial T_{n,p}}\right) + P_n\right]\nabla \cdot \mathbf{J}_n - T_L\left[\left(\frac{\partial \phi_p}{\partial T_{n,p}}\right) + P_p\right]\nabla \cdot \mathbf{J}_p$$

In equation (2.4), $\mathbf{J}_n$ and $\mathbf{J}_p$ are the non-isothermal current densities, of the form:

$$\mathbf{J}_n = -q\mu_n n\left(\nabla \phi_n + P_n \nabla T_L\right)$$
$$\mathbf{J}_p = -q\mu_p p\left(\nabla \phi_p + P_p \nabla T_L\right),$$

and P_n and P_p are absolute thermoelectric powers for electrons and holes and are calculated using:

$$P_n = -\frac{k_B}{Q}\left(\frac{5}{2} + \ln\left(\frac{N_c}{n}\right) + KSN + \zeta_n\right)$$

$$P_p = \frac{k_B}{Q}\left(\frac{5}{2} + \ln\left(\frac{N_v}{p}\right) + KSP + \zeta_p\right) \qquad (2.5)$$

In the last expression, k_B is the Boltzmann constant, KSN and KSP are the exponents in the power law relationship between relaxation time (or mobility) and carrier temperature. Their values typically range between -1 and 2 in silicon. The quantities ζ_n and ζ_p are the phonon drag contribution to the thermopower. These are only significant in low doped material and at low temperatures. The Silvaco Atlas module has a built-in model for this phonon drag contribution [8].

Case 2: Within the electron-lattice scattering model, the thermal system is represented as a single lattice temperature and is considered to be in thermal equilibrium. However, since the heat generation is due to a non-equilibrium electron population characterized by an electron temperature, T_e, the source term is then taken as a scattering term obtained from the relaxation time approximation and moments of the BTE. In essence, the transport is similar to case 1 in that the heat diffusion equation governs transport in the solid, except for the fact that the source term is now given as a moment of the relaxation time approximation, i.e.

$$H = \frac{3\rho k_B}{2}\left(\frac{T_e - T_L}{\tau_{e-L}}\right). \qquad (2.6)$$

In the above expression, T_e is the electron temperature, T_L is the lattice temperature and τ_{e-L} is the electron lattice energy relaxation time constant.

Case 3: Phonon-model. Under thermal non-equilibrium conditions a system of two phonons is used as represented in equations (2.14) and (2.15) later in the text. In this case, the 'lattice' temperature is taken to be the acoustic phonon temperature T_A, because this is the mode responsible for energy diffusion.

The energy balance equations for the acoustic and optical modes were for the first time derived by Majumdar and co-workers starting from the phonon Boltzmann transport equation [1]. In all the other approaches reviewed in this chapter, we have utilized this approach for the description of the phonon bath. A variant of this approach, that has been pursued by the Leeds group [9] and by Pop and co-workers [8], counts the number of generated acoustic and optical phonons in a given branch and mode. Then, the total heat generation rate per unit volume is computed as:

$$H = \frac{n}{N_{sim}\Delta t} \sum (\hbar\omega_{ems} - \hbar\omega_{abs}) \tag{2.7}$$

where n is the electron density, N_{sim} is the number of simulated particles and Δt is the simulation time.

One of the best solutions to the submicron heat transport problem by itself has been provided by Narumanchi and co-workers [10]. In their study they proposed a model based on the solution of the BTE, accounting for the transverse acoustic and longitudinal acoustic as well as optical phonons. Their model incorporates realistic phonon dispersion curves for silicon. The interactions among the different phonon branches and different phonon frequencies are considered and the proposed model satisfies energy conservation. Frequency-dependent relaxation times, obtained from perturbation theory, and accounting for phonon interaction rules, are used. In the calculation of the relaxation rates, they have included impurity scattering and the three-phonon interactions (the normal (N) process and the Umklapp (U) process). U processes pose direct thermal resistance while N processes influence the thermal resistance by altering the phonon distribution function. In this study, the BTE is numerically solved using a structured finite volume approach. Using this model, experimental in-plane thermal conductivity data for silicon thin films over a wide range of temperatures are matched in a satisfactory manner.

2.2 Solving lattice heating problem in nanoscale devices

Starting from the principle of energy conservation, Majumder and co-workers derived separate energy balance equations for the optical and acoustic phonon bath [4, 11]. Under the application of electric fields greater than 10 kV cm^{-1}, electrons tend to lose energy primarily to optical phonons; optical phonons decay further into acoustic phonons. The energy conservation equations for optical and acoustic phonons are

$$\frac{\partial W_{LO}}{\partial t} = \left(\frac{\partial W_e}{\partial t}\right)_{coll} - \left(\frac{\partial W_{LO}}{\partial t}\right)_{coll} \tag{2.8}$$

$$\frac{\partial W_A}{\partial t} = \nabla \cdot (\kappa_A \nabla T_A) + \left(\frac{\partial W_{LO}}{\partial t}\right)_{coll} + \left(\frac{\partial W_e}{\partial t}\right)_{coll} \tag{2.9}$$

where W_e, W_{LO} and W_A are electron, optical phonon and acoustic phonon energy densities, respectively. Since

$$dW_{LO} = C_{LO}\, dT_{LO}, \tag{2.10}$$

and

$$dW_A = C_A\, dT_A, \tag{2.11}$$

where C_{LO} (specific heat capacity for optical phonons) can be estimated using the Einstein model and C_A (specific heat capacity for acoustic phonons) from the Debye model. Next, the collision terms are expressed using the relaxation time approximation (RTA):

$$\left(\frac{\partial W_e}{\partial t}\right)_{coll} = n\frac{\frac{3}{2}k_B T_e + \frac{1}{2}m^* v_d^2 - \frac{3}{2}k_B T_{ph}}{\tau_{e-ph}} \tag{2.12}$$

$$\left(\frac{\partial W_{LO}}{\partial t}\right)_{coll} = C_{LO}\frac{T_{LO} - T_A}{\tau_{LO-A}} \tag{2.13}$$

where T_e is the electron temperature, v_d is the electron drift velocity and T_{ph} can be the optical or acoustic phonon temperature, depending on which kind of phonons the electrons interact with. Combining equations (2.8) and (2.13), one arrives at

$$C_{LO}\frac{\partial T_{LO}}{\partial t} = \frac{3}{2}nk_B\left(\frac{T_e - T_{LO}}{\tau_{e-LO}}\right) + \frac{nm^* v_d^2}{2\tau_{e-LO}} - C_{LO}\left(\frac{T_{LO} - T_A}{\tau_{LO-A}}\right) \tag{2.14}$$

$$C_A\frac{\partial T_A}{\partial t} = \nabla(\kappa_A \nabla T_A) + C_{LO}\left(\frac{T_{LO} - T_A}{\tau_{LO-A}}\right) + \frac{3}{2}\frac{nk_B}{\tau_{e-A}}(T_e - T_A) + \frac{1}{2}\frac{nm^* v_d^2}{\tau_{e-A}}. \tag{2.15}$$

The first two terms on the right-hand side (RHS) of (2.14) represent the energy gained from the electrons, where n is the electron density, T_e is the electron temperature, v_d is the drift velocity and T_{LO} is the optical phonon temperature, while the last term is the energy lost to the acoustic phonons. The same term also appears as a gain term on the RHS of equation (2.15). The first term on the RHS of equation (2.15) describes heat diffusion.

For the case where the electric fields are less than 10 kV cm^{-1}, electrons lose energy directly to the acoustic phonons and in that case, the energy balance equations can be expressed as:

$$C_A\frac{\partial T_A}{\partial t} = \nabla \cdot (\kappa_A \nabla T_A) + \left(\frac{\partial W_e}{\partial t}\right)_{coll} \tag{2.16}$$

$$C_A\frac{\partial T_A}{\partial t} = \nabla \cdot (\kappa_A \nabla T_A) - \frac{3}{2}\frac{nk_B}{\tau_{e-A}}T_A + n \cdot \frac{\frac{3}{2}k_B T_e + \frac{1}{2}m^* v_d^2}{\tau_{e-A}}. \tag{2.17}$$

Under the assumption of very low electric fields, the electron temperature and the acoustic phonon temperatures equal the lattice temperature, hence, the second and third terms in equation (2.17) cancel. Using the low field conductivity and the mobility expressions, the heat source term in equation (2.1) reduces to the last term of equation (2.17) as below

$$q_{gen} = \mathbf{J} \cdot \mathbf{E} = \sigma E^2 = \frac{\sigma v_d^2}{\mu^2} = \frac{nm^* v_d^2}{\tau}. \tag{2.18}$$

It is assumed here that for low doping concentrations, the relaxation time, τ in equation (2.18) is the acoustic phonon relaxation time. The reason for this assumption is that the acoustic phonon scattering process, being isotropic in nature, is mostly effective in randomizing the carrier momentum while the carrier energy is very low (under the application of low electric fields). The local Joule heating approximation given by the result in equation (2.18) is only valid for low fields, which is not the case in nanoscale devices.

In the following, we first give in section 2.2.1 a brief description of ASU's electron Monte Carlo device simulator, more details of which can be found in [12]. We then elaborate the coupling of the electron particle-based device simulator with the energy balance equations for phonons (section 2.2.2). Simulation results and discussion about the role of velocity overshoot and the influence the gate electrode temperature on the on-current in a nanoscale FD-SOI device with gate oxide, as well as with gate stack (high-k) are presented in sections 2.2.3–2.2.5, respectively. In section 2.2.6 we examine thermal degradation with scaling of FD-SOI device geometry. Alternative materials for the buried layer in SOI devices, such as diamond or aluminum nitride, and how they can improve the heat removal from the devices are presented in section 2.2.7. To account for the heating at the source, gate and drain interconnects, a novel multi-scale simulation approach, that combines circuit level simulations with device level simulations has been proposed. The approach compares simulation results with experimental measurements in an attempt to uncover the temperature profile due to self-heating effects. This is a subject of interest in section 2.3

2.2.1 Electron Monte Carlo for Si

In the theoretical model for the electron Monte Carlo (MC) we include all relevant scattering mechanisms for electrons in Si, including Coulomb scattering, acoustic deformation potential and intervalley g- and f-scattering processes. We assume general non-parabolic bands for the electrons and assume isotropic dispersions for the acoustic and optical phonons. The phonon dispersion relations in the reduced zone representation are shown in figure 2.1. It is assumed that the phonon dispersions are almost spherically symmetric.

The average electron velocities for $T = 77$ K, 200 K, 300 K and 430 K as a function of the electric field are shown in figure 2.2. Also shown here are the Pop *et al* [14] simulation data and experimental data from the literature [15].

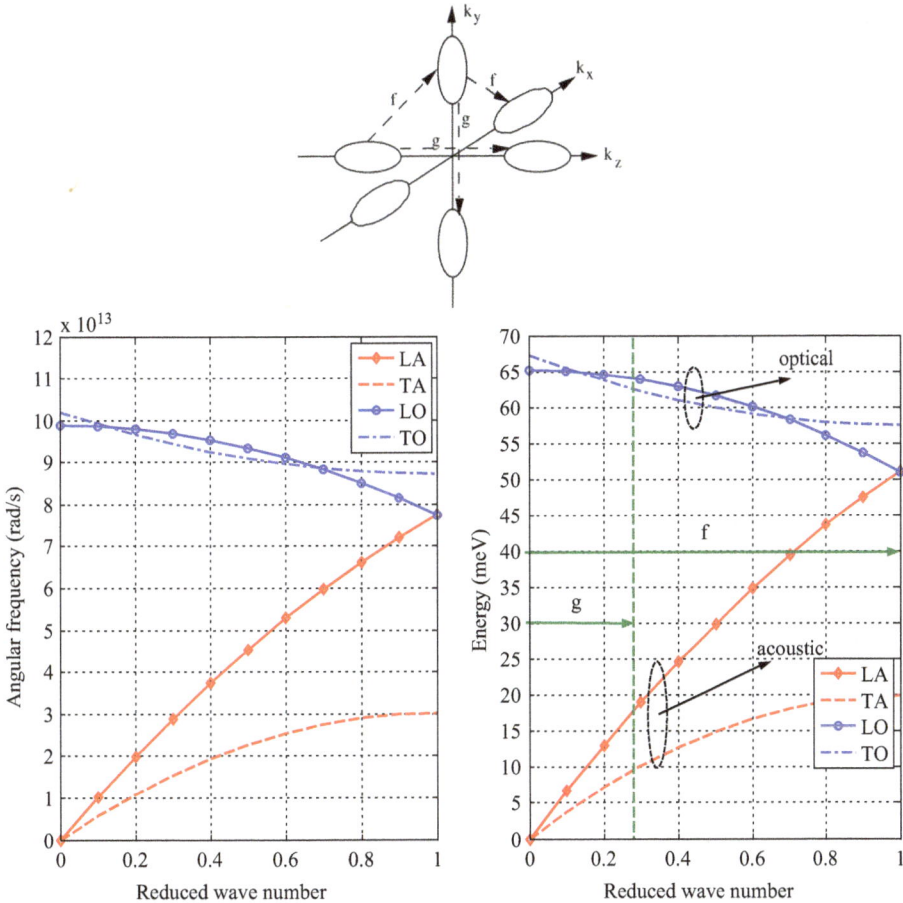

Figure 2.1. Description of the *g*- and *f*-phonons included in the theoretical model for Si (top panel) and their corresponding dispersion relations (bottom panel). Courtesy of Eric Pop [2].

Excellent agreement between the three sets of data is observed for the temperature range considered in this study. The accuracy of the scheme is further justified by the electron mobility plots versus electric field shown in figure 2.3 for $T = 77$ K, 200 K, 300 K and 430 K. Again, excellent agreement is observed when compared to the simulation data of Pop *et al* [3] and the experimental data [16].

2.2.2 Coupling of the electron particle-based device simulator with the energy balance solver for phonons

To be able to study self-heating effects, the EMC (ensemble MC) code for the carrier BTE solution [18, 19] has to be modified. As there is variable lattice temperature in the hot-spot regions, one has to introduce *temperature dependent scattering tables*. For each combination of acoustic and optical phonon temperature, one energy dependent scattering table is created. These scattering tables involve additional steps

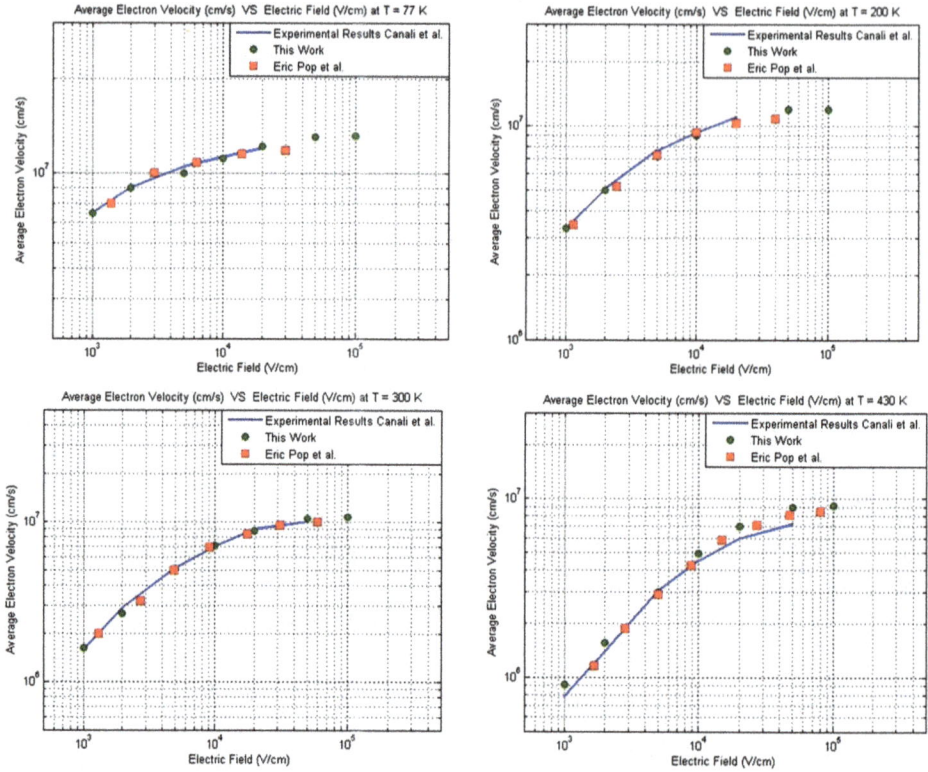

Figure 2.2. Average electron velocities in bulk silicon as a function of the electric field [13].

in the MC phase (figure 2.4—top right panel), because to choose randomly a scattering mechanism for a given electron energy, it is necessary to find the corresponding scattering table. To do that, first, the electron position on the grid needs to be found, in order to know the acoustic and optical phonon temperatures in that grid point, and then the scattering table with 'coordinates' (T_L, T_{LO}) is selected (figure 2.4—bottom panel). Using current state-of-the-art computers, the pre-calculation of these scattering tables does not require much CPU time or memory resources and has to be done once in the initialization stages of the simulation for a range of temperatures. An interpolation scheme is then adopted afterwards for temperatures for which an appropriate scattering table does not exist.

To properly connect the particle-based picture of electron transport with continuous, 'fluid-like' phonon energy balance equations, a space-time averaging and smoothing of the electron density, drift velocity and electron energy is included. At the end of each MC time step, the electrons are assigned to the nearest grid point. Then, the drift velocities and thermal energies are averaged with the number of electrons at the corresponding grid points. After the MC phase, a time averaging of the electron density, drift velocity and thermal energy is performed, and the electron temperature distribution is calculated. It is assumed that the drift energy is much

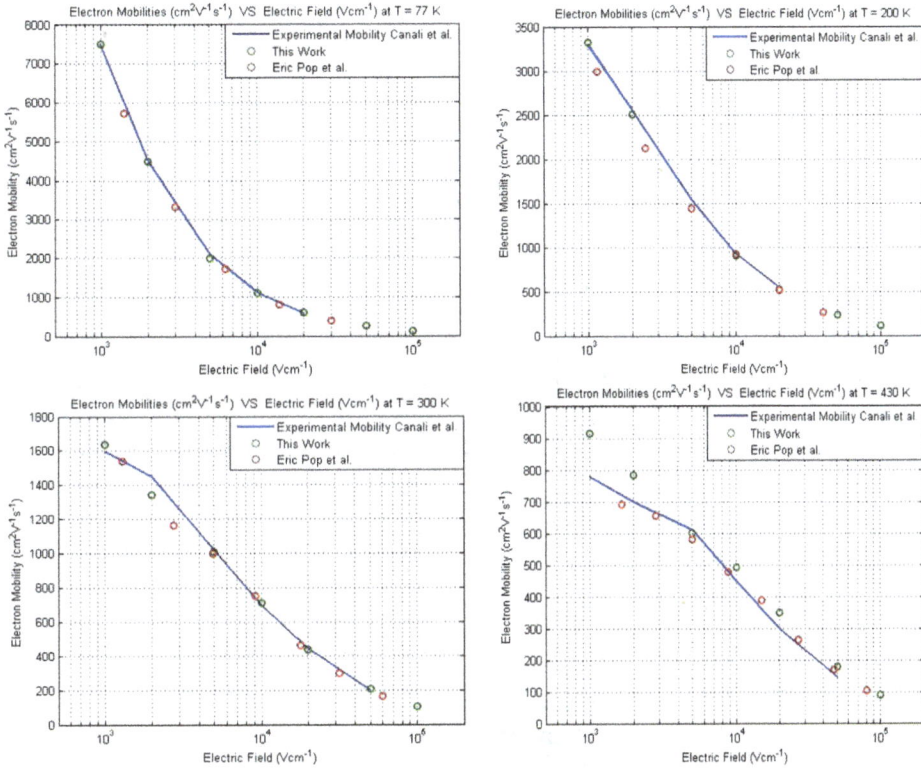

Figure 2.3. Low-field electron mobility in bulk silicon as a function of the electric field [17].

smaller than the thermal energy. The smoothing of these variables is necessary, because most of the grid points, especially at the interfaces, are rarely populated with electrons. This leads to very low lattice temperatures in those points. The exchange of variables between electron and phonon solvers is shown on the top left panel of figure 2.4.

Since SOI devices consist of two distinct regions, the silicon device layer and the buried oxide layer (in which the phonons have significantly smaller mean-free paths), the phonon BTE is solved in the silicon layer to accurately model heat transport, but the simpler heat diffusion equation is used in the amorphous BOX because the characteristic length-scale of heat conduction is much smaller than the film thickness. The two distinct computational regions are coupled through interface conditions that account for differences in material properties. For the coupling of the silicon and oxide solution domains, it is necessary to calculate the flux of energy through the interface between the two materials at each point along the interface for every time step.

The boundary conditions used have been chosen based on those typically used in commercial simulators. The Silvaco ATLAS simulation package [8] (THERMAL3D module) states that the only thermal contact should be the substrate. Also, according

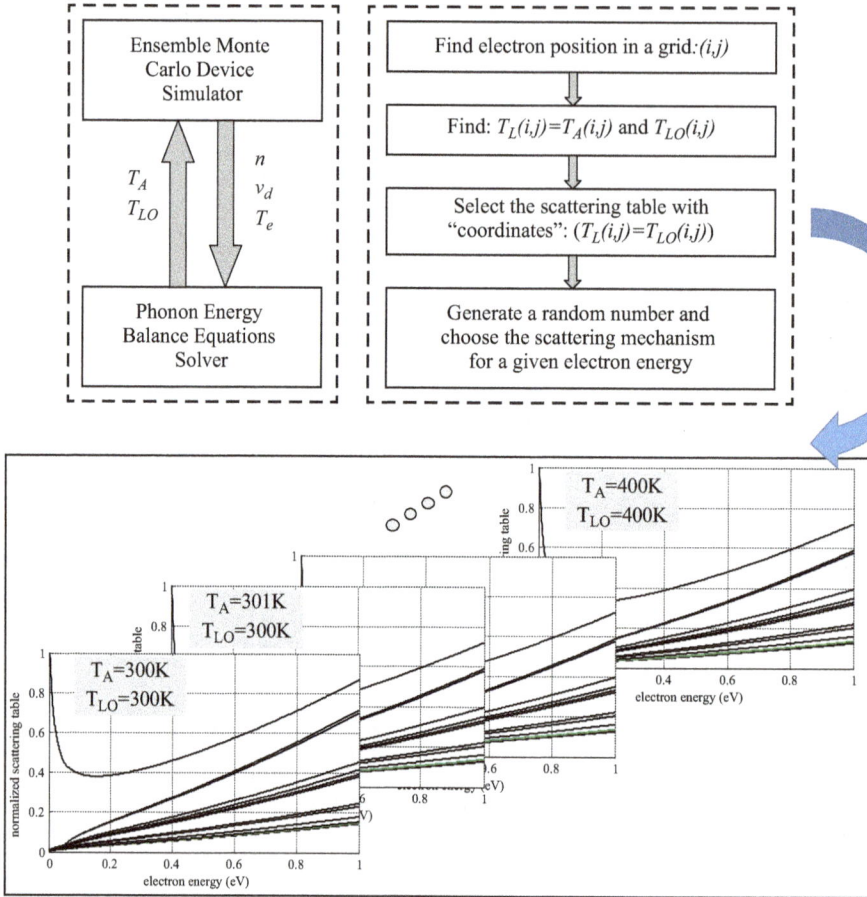

Figure 2.4. Top left panel: exchange of variables between the two kernels. Top right panel: choice of the proper scattering table. Bottom: set of temperature dependent scattering tables (normalized) as a function of electron energy—generated in the beginning of the MC phase.

to prescriptions given in the Silvaco ATLAS package, the source and drain should be left floating and the only electrode where one could specify isothermal boundary conditions is the gate. Since current nanoscale devices use metal gates to avoid poly-silicon depletion, such an isothermal assumption is seemingly justified. However, to study the efficacy of the gate as a heat sink, one has to simulate the effect on the current of several different temperatures for the gate, or use constant heat flux as a boundary condition.

2.2.3 Self-heating and the role of velocity overshoot

The thermal simulator briefly described in section 2.2.2 is used in the investigation of the role of heating effects on the electrical characteristics (on-state current) in different generations of nano-scale fully-depleted (FD) silicon-on-insulator (SOI) devices. Both standard SiO_2 gate oxide and gate-stack ($HfO_2 + SiO_2$) dielectrics in the smallest structures being investigated, are considered where high-k dielectrics are used to prevent high off-state leakage currents. Contrary to common expectations, it was found that the degradation of the current decreases with decreasing device gate length due to the more pronounced velocity overshoot effect in the device characteristics which results in smaller transit time of the carriers through the active region which, in turn, reduces the probability for interaction with the phonons. This observation is clearly seen from the results presented below. The dimensions of the n-channel fully-depleted (FD) SOI MOSFET being investigated are: the channel length is 25 nm, the silicon film thickness, which is equal to the source/drain junction depth is 10 nm, the gate oxide thickness is 2 nm, the BOX thickness is 50 nm, the source/drain doping is 1×10^{19} cm^{-3} and the channel doping is 1×10^{18} cm^{-3}.

To simulate the steady-state state behavior of a device, the system is started in some initial condition, with the desired potential applied to the contacts, and then the simulation proceeds in a time stepping manner until steady-state is reached. A common starting point for the initial guess is to start out with charge neutrality, i.e., to assign particles randomly according to the doping profile in the device, so that initially the system is charge neutral on the average. After assigning charges randomly in the device structure, charge is assigned to each mesh point using an adequate PM (particle-mesh) coupling method, and the 2D Poisson equation is solved. The forces are then interpolated on the grid, and particles are accelerated over the next time step (see figure 2.5). When the system is driven into a steady-state regime and MC simulation time has elapsed, the steady-state current through a specified terminal is calculated.

To continue with the thermal part of the simulation, the average electron density, drift velocity and electron temperature must be calculated on a grid. For the given

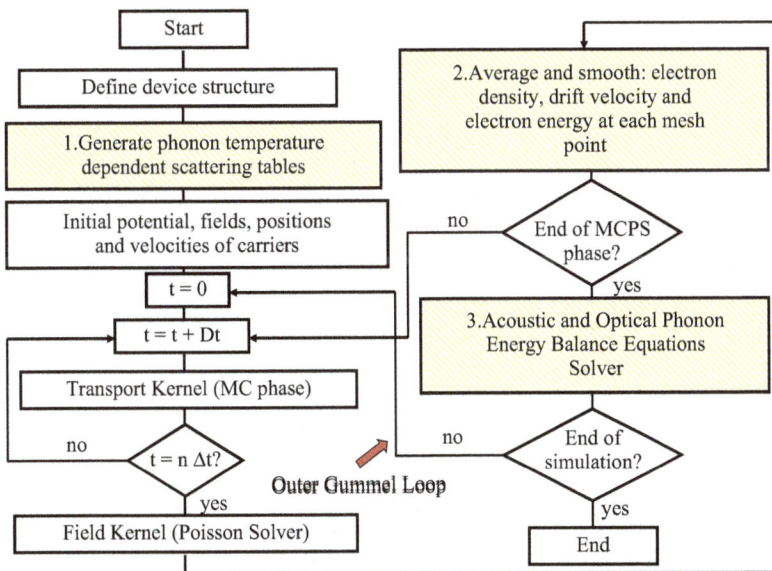

Figure 2.5. Flowchart of the electro-thermal particle based device simulator.

x 10^{25}

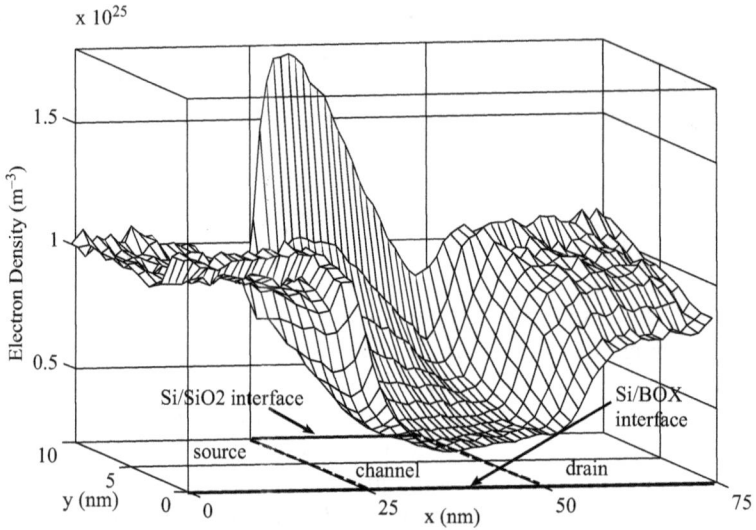

Figure 2.6. Sample electron density [12].

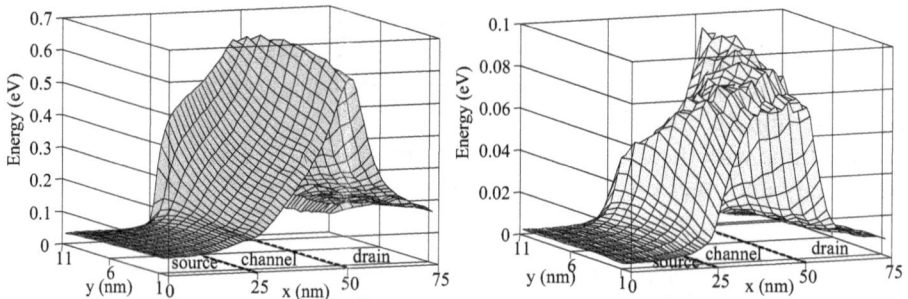

Figure 2.7. Left panel: electron thermal energy. Right panel: electron drift energy [12].

bias conditions, the average electron density in the channel (figure 2.6) is very high at the Si–SiO$_2$ interface near the source injection barrier, while pinch off region exists near the drain.

The two components of the electron kinetic energy are presented in figure 2.7. They show that the thermal part of electron kinetic energy is much larger than the drift part, so electron temperature can be calculated from the thermal energy.

Finally, the acoustic and optical phonon temperatures are calculated with the phonon energy balance equation solver. During the simulation, the gate contact and the bottom of the BOX is set to 300 K, while Neumann boundary conditions for the heat transfer are used in all other outer surfaces. Also, the tolerance used in the 'thermal' SOR algorithm equals 0.001, which leads to very fast convergence.

When the simulation starts, all variables obtained from the first iteration of the EMC solver, are calculated using a uniform distribution for the acoustic and optical phonon temperatures. This means that only one scattering table is used for all

Table 2.1. Percentage of different electron–phonon interactions with isothermal and thermal device simulator.

Intervalley Scattering:	Number of events (isothermal simulation)	(%) (isothermal simulation)	Number of events (thermal simulation)	(%) (thermal simulation)
Acoustic phonon (absorption)	4234	9.38	17 425	10.5
Acoustic phonon (emission)	7863	17.42	27 864	16.8
Optical phonon (absorption)	4017	8.9	22 202	13.8
Optical phonon (emission)	29 020	64.3	98 408	59.32

electrons, no matter where they are located in the device. When the phonon temperatures are computed from the phonon energy balance equations, they are 'returned' at the beginning of the MC free-flight-scattering phase. Now, for each mesh point, one has a scattering table which corresponds to the acoustic and optical phonon temperatures at that point. In this case, the electron position defines which scattering table is 'valid' and then, by generating a random number, the scattering mechanism is chosen for the given electron energy. The impact of the phonon temperature dependent scattering tables can be demonstrated by counting the number of energy-exchange electron–phonon scattering events. From table 2.1, it could be concluded that the inclusion of the phonon temperature dependent scattering table increases the number of electron–phonon interactions. It was found that electrons with energies below 50 meV scatter mainly with acoustic phonons in silicon, while those with higher energy scatter strongly with the optical modes. The optical phonon modes have low group velocity (on the order of 1000 m s^{-1}) and their occupation is also relatively low, hence they contribute very little to the heat transport. The primary heat carriers in silicon are the faster acoustic phonon modes, which are significantly populated and have group velocities from 5000 m s^{-1} for transverse modes to 9000 m s^{-1} for longitudinal acoustic modes. Optical phonons decay into acoustic modes, but over relatively long time scales, i.e. picoseconds, compared to the order of tenths of picoseconds [20]. Under high field conditions, this can lead to the creation of a phonon energy bottleneck which can cause the density of optical phonon modes to build up over time, leading to more scattering events and impeding electron transport [21]. This is clearly seen from the results shown in table 2.1.

The outer Gummel loop between the MC solver and the phonon energy balance solver (see figure 2.5) ends when the steady-state conditions for the phonon temperatures and the device current are reached. To test the overall convergence of the coupled thermal and EMC codes, the *variation* of the drain current with the number of thermal iterations for a given bias condition are recorded. The results of

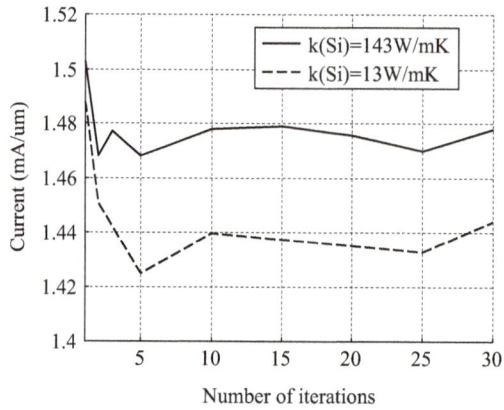

Figure 2.8. Current convergence [12].

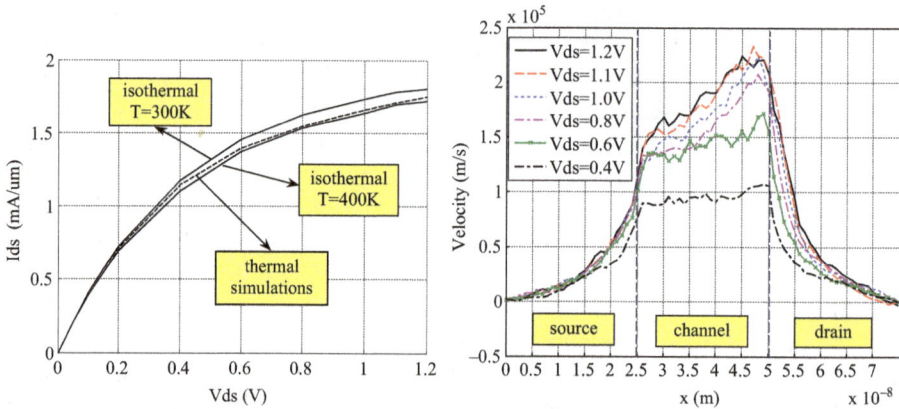

Figure 2.9. Left panel: output characteristics for $V_{GS} = 1.2$ V. Top/middle curve corresponds to the case of excluded/included lattice heating model. The bottom curve is an isothermal simulation but for lattice temperature $T = 400$ K throughout the whole simulation domain. Right panel: velocity along the channel for $V_{GS} = 1.2$ V and different values of V_{DS}. Note that for $V_{DS} > 0.4$ V electrons are in the velocity overshoot regime, which suggests that lattice heating does not significantly degrade the device characteristics [12].

these simulations (figure 2.8) show that only 3–5 thermal iterations are necessary to obtain steady-state solution of the current. Also, when thermal simulations are self-consistently coupled with the EMC code, a smaller number of time steps are needed for obtaining a steady-state condition in the EMC phase. For example, to get equal slopes from the cumulative source and drain charge time characteristics, the simulation time for isothermal EMC code should be at least 6 ps, while 2.5 ps simulation time for MCPS phase is enough to reach steady-state when thermal simulations are included.

The left panel of figure 2.9 shows the effect of lattice heating on the *I–V* characteristics of a 25 nm gate length fully depleted SOI structure. The velocity

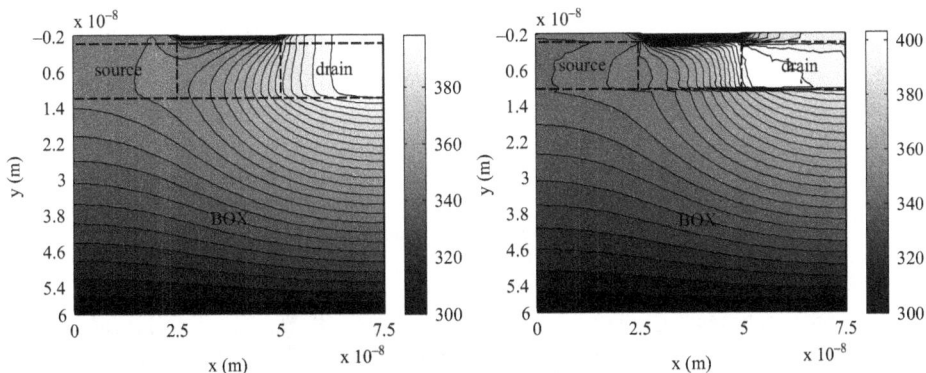

Figure 2.10. Left panel: acoustic phonon temperature for $V_{GS} = 1.2$ and $V_{DS} = 1$ V. Notice the significant heating of the lattice that equals the acoustic phonon temperature in our model. Right panel: optical phonon temperature for $V_{GS} = 1.2$ V and $V_{DS} = 1$ V. Note the region near the drain with higher optical phonon temperature with respect to the acoustic phonon temperature [12].

Table 2.2. Current variation with gate temperature for 25 nm fully-depleted SOI device structure with SiO_2 as gate oxide. Bias conditions: $V_{GS} = V_{DS} = 1.1$ V.

Type of simulation	Gate Temperature	Current Decrease
thermal	300 K	5.1%
thermal	400 K	9.18%
thermal	600 K	17.12%

overshoot is clearly seen in the right panel of figure 2.9. For this particular device structure, the corresponding degradation of device characteristics due to thermal effects is relatively small, less than 10%. As seen from the temperature maps of acoustic and optical phonons in figure 2.10, the maximum rise in the lattice temperature is on the order of 100 K on the drain side of the gate as expected. For comparison, we also compare in figure 2.9 the effect of an elevated lattice temperature of 400 K assuming an isothermal model compared to the non-uniform model, which shows that most of the effect observed in the I–V characteristics is due to lattice heating.

2.2.4 Influence of the gate electrode boundary condition on the on-current

As already discussed in section 1.2, to properly solve the phonon balance equations, the device should be attached to a heat sink somewhere along the boundary or finite heat conduction through the surface should be allowed for. In the present model, a heat sink is modeled by a simple Dirichlet boundary condition (i.e. constant temperature). The gate electrode contact and the bottom of the BOX is used as a heat sink. Table 2.2 gives the percentage of the current decrease due to heating

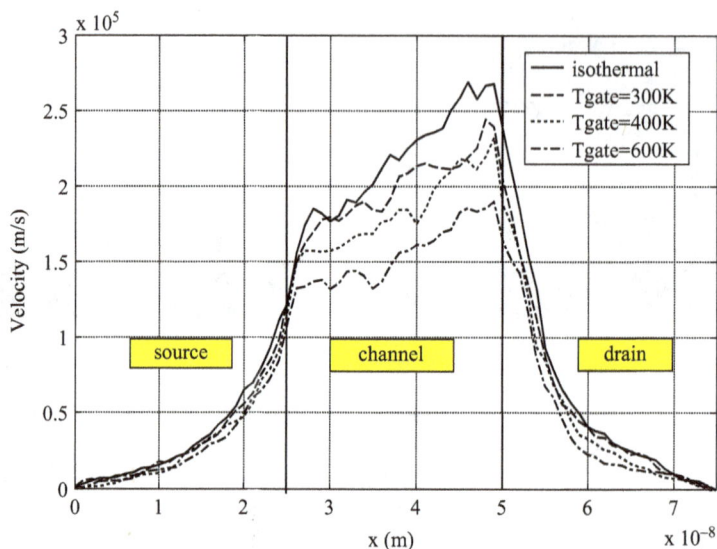

Figure 2.11. Velocity along the channel for $V_{GS} = 1.1$ V and $V_{DS} = 1.1$ V for different gate temperatures. Note that the electrons in the channel are in the velocity overshoot regime (electron saturation velocity in silicon is 1.1×10^5 m s^{-1}) [12].

effects with the variation of the gate electrode temperature. The calculated results show that the current degradation is more prominent for higher gate temperatures. When the temperature of the bottom of the BOX is set to the same values as given in table 2.2, the current degradation is around 1%, so in all other simulations the temperature of the bottom of the BOX is set to 300 K.

Figure 2.11 shows the velocity profile along the channel for the same bias conditions ($V_{GS} = V_{DS} = 1.1$ V) and different gate temperatures, where, as can be seen, the velocity in the channel decreases with the increase of the gate temperature, but the carriers in the channel are still in the velocity overshoot regime. As seen from the temperature maps of acoustic phonons in figure 2.12, the lattice temperature in the source, the channel and the drain region is increasing with the increase of the gate electrode temperature, which means that the increased lattice temperature has a larger impact on the decrease of the carrier velocity in the channel.

2.2.5 Thermal degradation in high-k devices

To further investigate the influence of the gate temperature on the current decrease due to the self-heating, instead of SiO_2 as the gate oxide, we used a gate stack (SiO_2 and HfO_2) and a copper metal gate with finite thickness for the same 25 nm fully-depleted SOI device structure. In order to compare the results of these two device structures, the gate stack thickness was calculated in such a manner that the surface potential for $V_{GS} = 1.1$ V and $V_{DS} = 0$ is the same for both. The results of the thermal simulations for the fully-depleted SOI structure with this gate stack are given in figures 2.13 and 2.14. When the gate electrode is modeled with zero thickness (see figure 2.13), the current decrease due to the thermal effects is 5% more

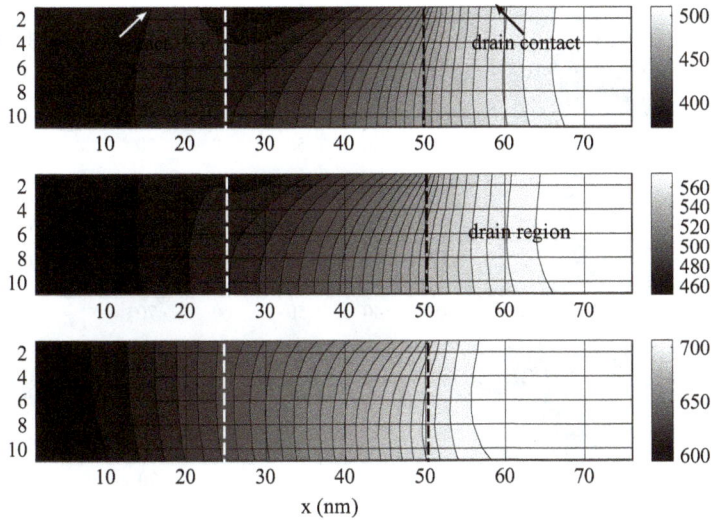

Figure 2.12. Lattice temperature profiles in the silicon layer in 25 nm gate-length fully-depleted SOI MOSFET for $V_{GS} = 1.1$ V and $V_{DS} = 1.1$ V and different gate electrode temperatures (300 K, 400 K and 600 K from up to down). SiO_2 is used as gate oxide [12].

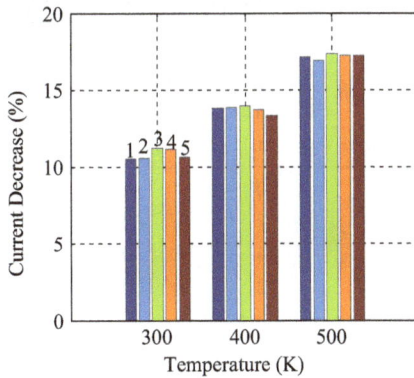

Figure 2.13. Current decrease with gate temperature for 25 nm fully-depleted SOI device with HfO_2 gate stack. Parameter is the metal gate electrode thickness (1–0 nm; 2–3 nm; 3–5 nm; 4–7 nm; 5–40 nm). The boundary condition for the temperature is at the end of the metal gate, which means that the thermal simulator also solves the metal region.

when a gate stack is used instead of SiO_2 as a gate dielectric. One can also note that the details of gate electrode thickness modeling have little effect on the simulated current degradation. The lattice temperature profile in the silicon layer for different gate temperatures, with 40 nm gate electrode thickness, is shown in figure 2.14. As can be seen from the results presented in this figure, the lattice temperature is increased due to the gate stack compared to the corresponding lattice temperature profile from figure 2.12.

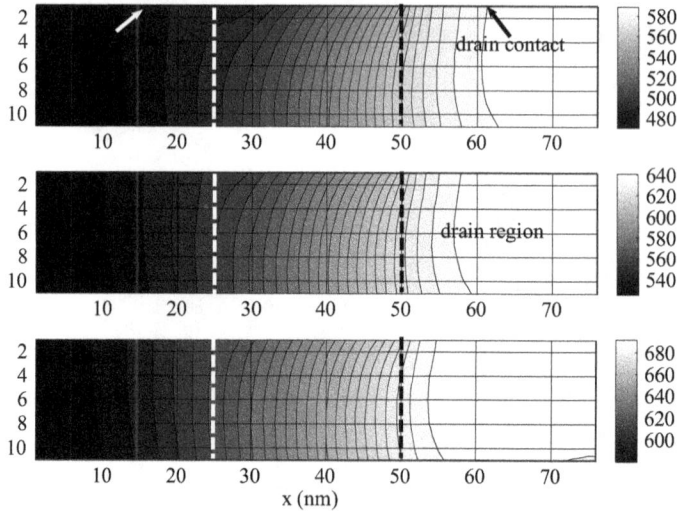

Figure 2.14. Lattice temperature profiles in the silicon layer in 25 nm gate-length fully-depleted SOI MOSFET for $V_{GS} = 1.1$ V and $V_{DS} = 1.1$ V and different gate temperatures (300 K, 400 K and 500 K from up to down). Gate stack (SiO_2 and HfO_2) is used [12, 22].

Table 2.3. Geometrical dimensions of the simulated fully-depleted SOI MOSFETs.

Gate length L_{gate} (nm)	Gate oxide thickness t_{OX} (nm)	Source/ drain length $L_{S/D}$ (nm)	Si-layer thickness t_{si} (nm)	BOX thickness t_{BOX} (nm)	Source/drain doping concentration N_D (m^{-3})	Channel doping concentration N_A (m^{-3})
25	2	25	10	50	1×10^{25}	1×10^{24}
45	2	45	18	60	1×10^{25}	1×10^{24}
60	2	60	24	80	1×10^{25}	1×10^{24}
80	2	80	32	100	1×10^{25}	1×10^{23}
90	2	90	36	120	1×10^{25}	1×10^{23}
100	2	100	40	140	1×10^{25}	1×10^{23}
120	3	120	48	160	1×10^{25}	1×10^{23}
140	3	140	56	180	1×10^{25}	1×10^{23}
180	3	180	72	200	1×10^{25}	1×10^{23}

2.2.6 Thermal degradation with scaling of device geometry

In addition to the previously noted observation regarding the influence of the velocity overshoot, we modeled larger fully-depleted SOI device structures and we also investigated the influence of the temperature boundary condition on the gate electrode on the current degradation due to self-heating effects. The geometrical dimensions of the simulated fully-depleted SOI MOSFETs are given in table 2.3, while figure 2.15 gives the percentage of the current decrease due to heating effects with the variation of the gate electrode temperature for the device structures being

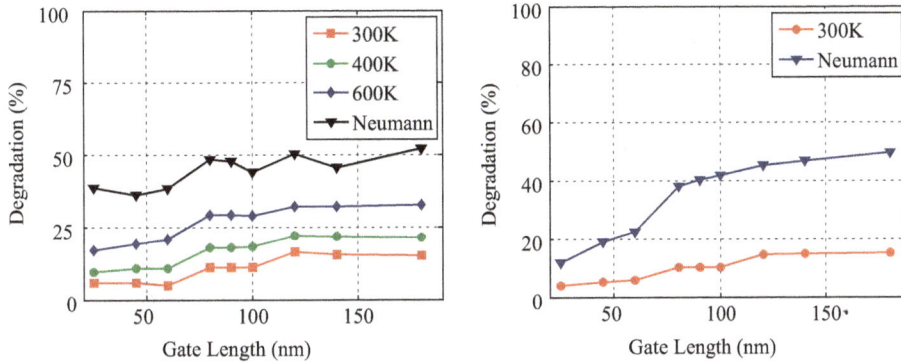

Figure 2.15. Current degradation versus technology generation ranging from 25 nm to 180 nm channel length FD SOI devices (see table 2.3). Isothermal boundary condition of 300 K is set on the bottom of the BOX. Parameter is the temperature on the gate electrode. Neumann (left panel)/Dirichlet (right panel) boundary conditions are applied at the vertical sides[1].

considered. Two types of simulation have been done: when thermal Neumann boundary conditions are applied at the vertical sides (which means no heat flow) (figure 2.15 left) and with thermal Dirichlet boundary conditions at the vertical sides with a constant temperature of 300 K (figure 2.15 right). The calculated results show that the current degradation is more prominent for larger devices and for higher gate temperatures. For 80 nm and larger devices, the simulated carriers are not in the velocity overshoot regime in the larger portion of the channel. Snapshots of the lattice temperature profiles in the silicon layer for these devices, when the gate temperature is set to 300 K and 400 K, are given in figures 2.16 and 2.17, respectively. From these snapshots one can observe that: (a) the temperature in the channel is increasing with the increase of the channel length, (b) the maximum lattice temperature region (hot spot) is in the drain and shifts towards the channel for larger devices. This behavior is more drastic for higher gate temperatures (see figure 2.17).

Figure 2.18 gives the ensemble averaged lattice and optical phonon temperatures along the channel in the silicon layer only for the three technologies of devices being considered (25 nm, 80 nm and 180 nm). Notice that there is a bottleneck between the lattice and the optical phonon temperature in the channel which is more pronounced for shorter devices, due to the fact that the energy transfer between optical and acoustic phonons is relatively slow compared to the electron-optical phonon processes and the fact that the electrons are in the velocity overshoot (and since the channel is very short, they spend little time in the channel). To better understand the phonon temperature bottleneck, different cross-sections (at the Si/SiO$_2$ interface, at the half Si-layer width, and at the Si/BOX interface) of the lattice and the optical phonon temperature profiles in the channel direction were investigated as well. We find that the bottleneck decreases from the Si/SiO$_2$ interface to the Si/BOX interface.

---

[1] The validity of the Dirichlet boundary condition at the bottom gate has been carefully checked by performing Silvaco THERMAL3D simulations on identical structures with and without the underlying silicon substrate. The difference in the results is less than 0.5%

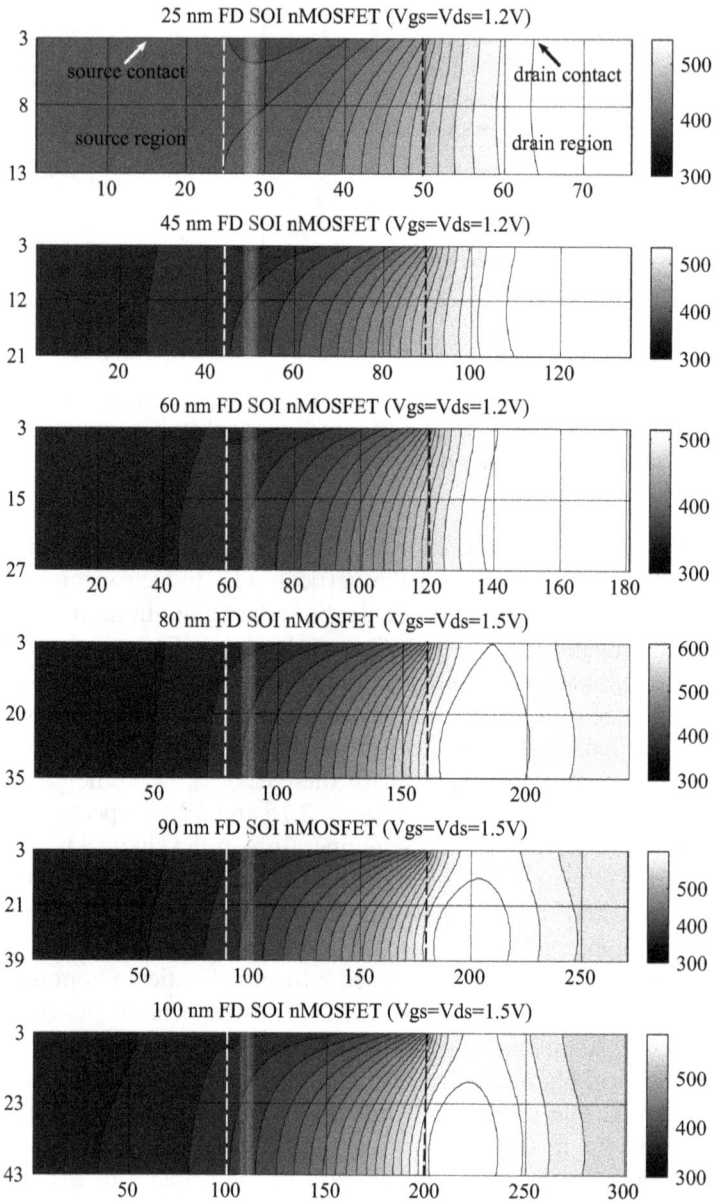

Figure 2.16. Lattice temperature profiles in the silicon layer for FD SOI MOSFETs from table 2.3 with gate temperature set to 300 K. (25 nm—top, 100 nm—bottom) [12, 22].

For shorter devices, it exists in the whole channel region, which is not a case for longer devices (thicker Si-layer and longer channel length). From the results we have presented here, one can conclude that the higher the temperature in the channel and/ or the longer the electrons are in the channel, the larger the degradation of the device electrical characteristics is due to the heating effects.

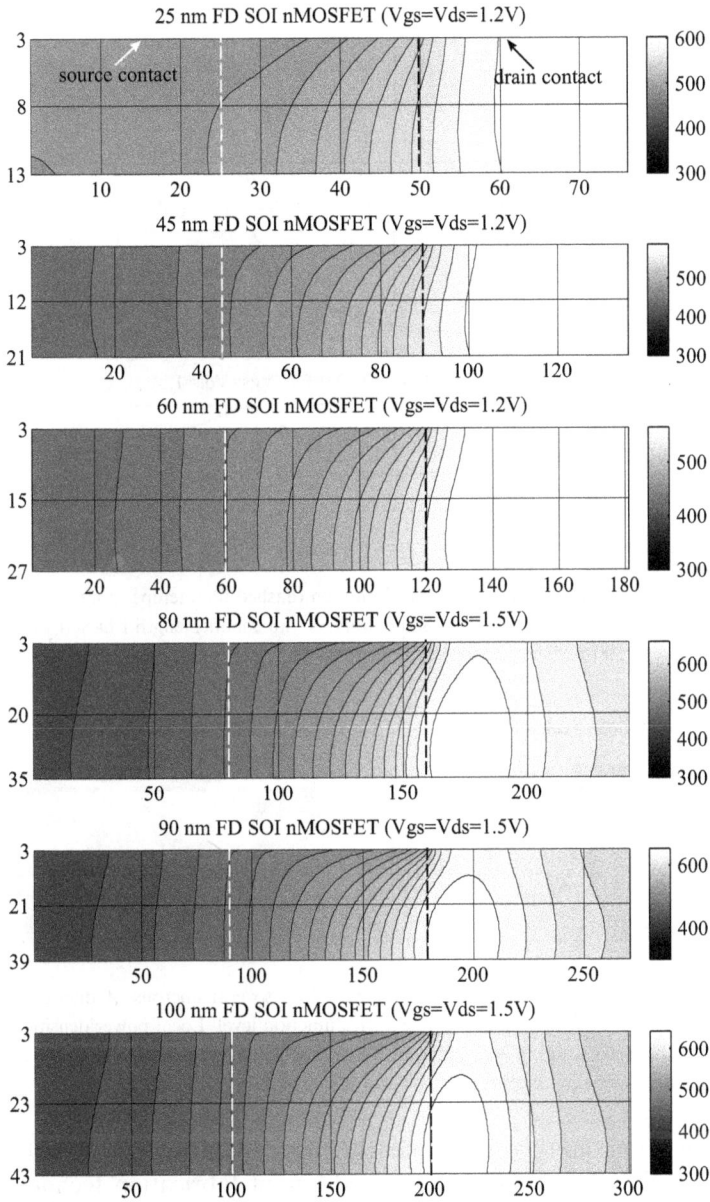

Figure 2.17. Lattice temperature profile in silicon layer for different device geometries of fully-depleted SOI MOSFET when gate temperature is set to 400 K [12, 22].

2.2.7 Modeling of SOD and SOAlN devices

One way of reducing self-heating in state-of-the art devices is to use a partial BOX layer, which is difficult to manufacture [23]. An alternative to this approach is to replace the buried silicon dioxide by another dielectric layer. The potential

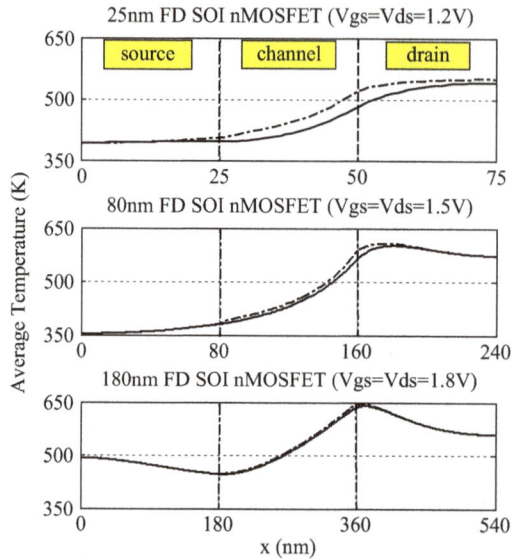

Figure 2.18. Averaged lattice (full line) and optical phonon (dashed line) temperature profiles in the channel direction in the active silicon layer for 25 nm, 80 nm and 180 nm channel length FD-SOI MOSFETs with the gate temperature set to 300 K. The phonon energy bottleneck is more pronounced for smaller devices.

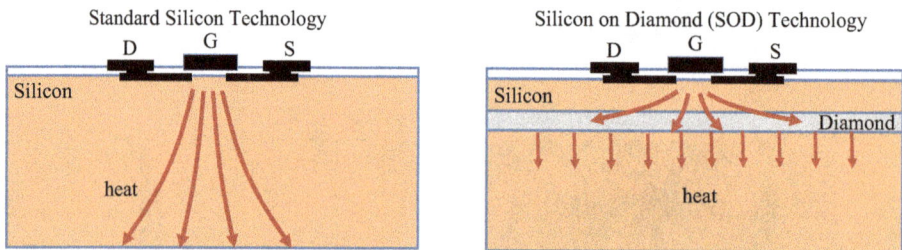

Figure 2.19. High diamond thermal conductivity means that several microns of diamond can effectively equalize temperatures across an entire chip at the device junction level. Local power densities of hundreds to thousands of watts per square mm in logic chips are spread quickly over the total area of the chip and reduced to a few hundred watts per square centimeter at most [25].

candidates must not only possess higher thermal conductance than SiO_2, but should also be compatible with current integrated circuit fabrication technology. In this regard, two materials are considered as potential substitutes [24]. Aluminum nitride (AlN), which has a thermal conductivity approximately 100 times higher than SiO_2 (136 W $(m{\cdot}K)^{-1}$ versus 1.4 W $(m{\cdot}K)^{-1}$) and roughly equal to that of silicon itself (145 W $(m{\cdot}K)^{-1}$), has excellent thermal stability, high electrical resistance and a coefficient of thermal expansion close to that of silicon. The second candidate is crystalline diamond with a thermal conductivity about 1000 times higher than that of SiO_2 (2000 W $(m{\cdot}K)^{-1}$ versus 1.4 W $(m{\cdot}K)^{-1}$). The diamond layer not only serves as a dielectric, but also as a heat diffuser, and should be as close to the heat generating region as is physically possible (see figure 2.19). The net result is to reduce both the local junction temperature as well as the overall chip temperature since the heat flow path is now the full area of the chip itself. The Si on diamond (SOD) structure is basically equivalent to an SOI device structure, but the insulator (I) in SOI is diamond rather than silicon dioxide.

One potential drawback regarding the substitution of buried SiO_2 with diamond or AlN is the quality of the bottom interface and the possibility of increased interface-roughness scattering that can lead to reduced current drive in FD-SOD devices. In the FD SOI/SOD and SOI/SOAlN device structures studied here, most of the carriers are traveling close to the top interface, so the contribution to the mobility due to the remote roughness scattering [26] of the back interface is relatively small. In addition, high resolution transmission electron micrograph studies [27] of the bonded Si/diamond interface, show that this interface can be made very smooth, which suggests that bonding techniques can minimize even the possibility of remote roughness scattering. Thus, it is very feasible that in the near future we may see SOD or SOAlN devices replacing conventional SOI devices.

In this section, we compare the current degradation in FD-SOD and FD-SOAlN devices to FD-SOI devices and also examine the nature of the heat flow in these device structures. In the simulations that follow, the quality of the silicon/diamond or silicon/AlN interface, for simplicity, is assumed to be equal to the quality of the top Si/SiO_2 interface in terms of surface roughness scattering. We have checked that by performing additional simulations in which the quality of the bottom interface was varied by changing the ratio of diffusive versus specular scattering.

Simulation results for a 25 nm FD-SOD(SOAlN) device are presented in figures 2.20 and 2.21, respectively. For these results, the only thermal boundary condition is the temperature of the substrate set to $T_{SUB} = 300$ K and the gate is left floating. Two things can be observed: (1) the hot-spot is moved more towards the channel when we compare the temperature maps for the lattice temperature in figures 2.20 and 2.21, and figure 2.10 for the FD-SOD(SOAlN) and FD-SOI device. (2) As in the

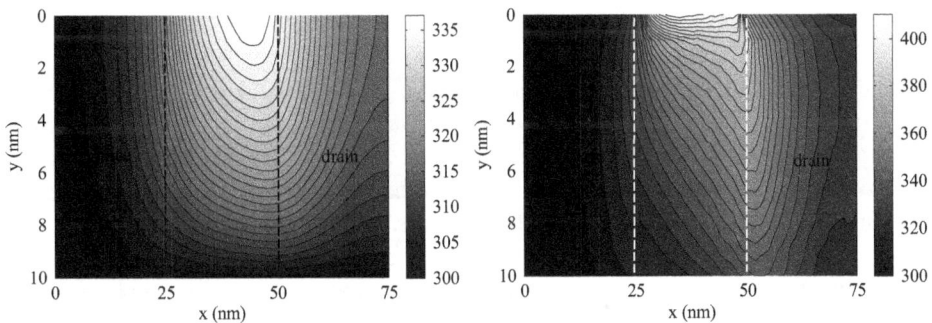

Figure 2.20. Left panel: lattice temperature profile for a 25 nm FD silicon-on-diamond MOSFET ($V_{GS} = V_{DS} = 1.2$ V). The Dirichlet boundary conditions for the thermal simulation is $T_{BOX} = 300$ K Right panel: Optical phonon temperature profile for 25 nm FD silicon-on-diamond MOSFET ($V_{GS} = V_{DS} = 1.2$ V). Dirichlet boundary condition for thermal simulation is $T_{SUB} = 300$ K [28]. Only the device active region is shown here [29].

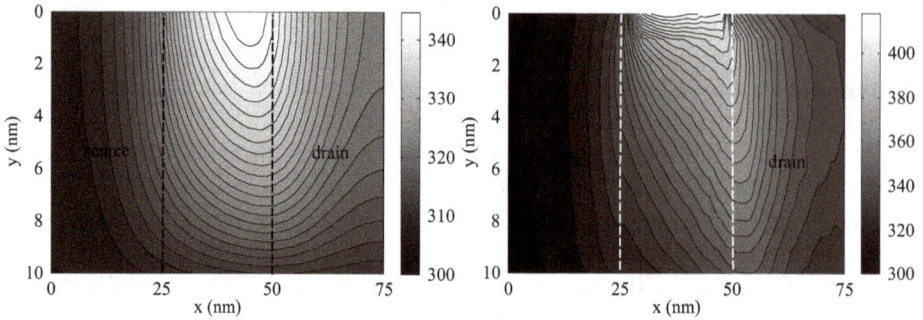

Figure 2.21. Left panel: lattice temperature profile for 25 nm FD silicon-on-AlN MOSFET ($V_{GS} = V_{DS} = 1.2$ V). Dirichlet boundary condition for thermal simulation: $T_{BOX} = 300$ K. Right panel: optical phonon temperature profile for 25 nm FD silicon-on-AlN MOSFET ($V_{GS} = V_{DS} = 1.2$ V). Dirichlet boundary condition for thermal simulation: $T_{SUB} = 300$ K. Only the device active region is shown here [29].

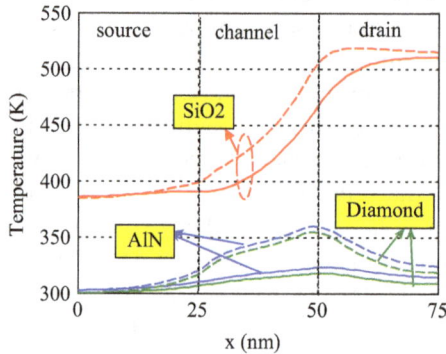

Figure 2.22. Average acoustic and optical phonon temperature profile in the active silicon layer for 25 nm channel-length fully-depleted SOI, SOD and SOAlN devices. Note that the phonon energy bottleneck is higher when SiO$_2$ is used as a BOX material.

case of FD-SOI devices, in FD-SOD(SOAlN) devices there is a bottleneck at the energy exchange between optical and acoustic phonons and the optical phonon's temperature is higher than the acoustic/lattice temperature. When we fix the temperature on the gate electrode to 300 K, the hot-spot moves even further towards the drain, but the bottleneck in the exchange of energy between the acoustic phonon and the optical phonon baths remains. As can be seen in figure 2.22, the spread of the temperature across the bottom side of the wafer is more uniform (which means that hot-spots are less likely to occur in these two technologies if adopted). There is slightly better heat spread in the SOD than in the SOAlN devices, which is in agreement with the results from [29].

The influence of the various boundary conditions on the gate electrode on the current degradation due to lattice heating is given in figure 2.23. We see maximum degradation of about 5% for SOD (SOAlN) and around 17% for FD-SOI devices

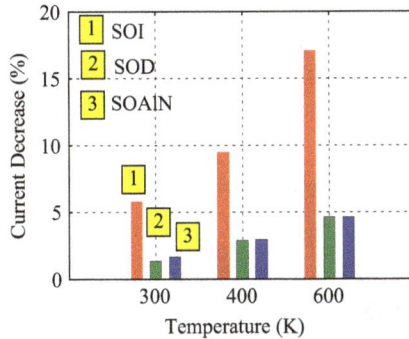

Figure 2.23. Current decrease due to a self-heating for 25 nm channel-length fully-depleted SOI, SOD and SOAlN devices for different gate temperature.

when the gate temperature is set to 600 K. One can conclude that for SOD (SOAlN) devices lattice heating plays a minor role in the current degradation.

We also found that the interplay of the thermal conductivity, thickness and dielectric constant for the buried insulator layer can considerably change the temperature maps and the amount of current degradation. In other words, given a buried material with specific values of thermal conductivity and the dielectric constant, there exists an optimum buried insulator thickness for which the degradation of the current characteristics is at minimum. Further details can be found in [30]. Issues that need to be investigated further include the role of remote roughness for thinner silicon films and shorter channel lengths, and the role of electron–electron interactions on the thermalization of the carriers in the drain end of the device.

2.3 Multi-scale modeling—modeling of circuits (CS and CD configuration)

International technology roadmap for semiconductors (ITRS) suggests that as devices are scaled to smaller dimensions, the current density in the interconnects would increase [31, 32]. Hence, it is important to account for heating effects not only within the device itself, but also the contacts and interconnects when considering the reliability of a system. To do exactly that, a novel multi-scale simulation approach, that combines circuit level simulations with device level simulations, has been proposed. The approach compares simulation results with experimental measurements in an attempt to uncover the temperature profile due to self-heating effects. The proposed method couples circuit level simulations performed using Silvaco Atlas [8] with an electro-thermal Monte Carlo device simulator [33]. The Giga3D Silvaco Atlas module simulates the thermal transport characteristics at the interconnect level. This module provides temperature boundary conditions for the device-level simulation. Then the device level simulator solves for self-heating throughout the device. The coupled system is shown in figure 2.24. It is important

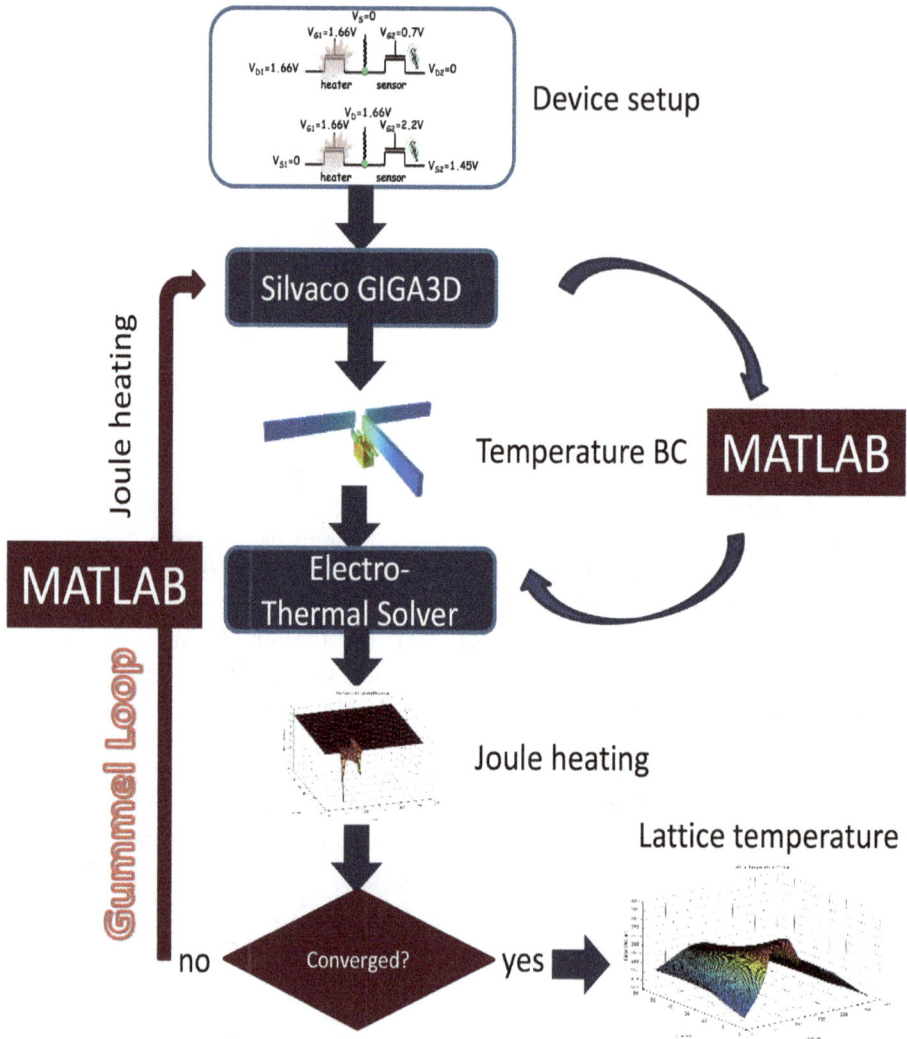

Figure 2.24. Implemented multi-scale modeling scheme.

to note that in the device level thermal solver, the 2D/3D Poisson equation is solved self-consistently with an MC transport kernel coupled to a 2D/3D energy balance equation solver for the acoustic (lattice) and optical phonon baths [34] as described in sections 2.2.1 and 2.2.2. This is, by itself, a new multi-scale approach which is very different from the commonly used Joule heating model used in commercial device simulators. Such simulations give rise to more pronounced hot-spots, because they accurately represent the optical-to-acoustic phonon bottleneck [35].

The device level temperature measurement technique used in this work is based on the temperature dependence of the sub-threshold slope of a planar MOSFET. The underlying idea is that variations in the sub-threshold slope can be used to determine the temperature within the hot-spot. Two devices in either common

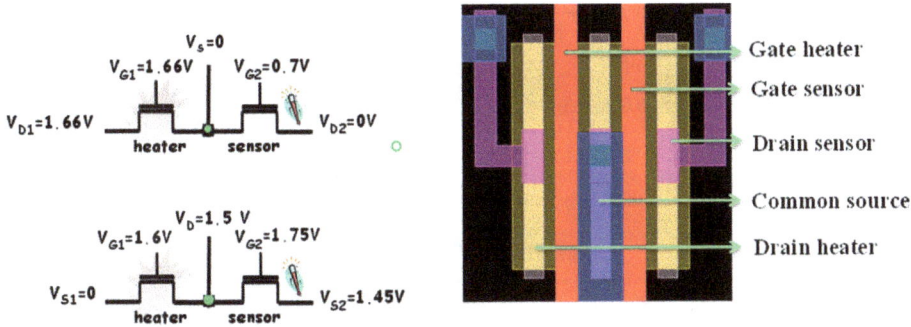

Figure 2.25. Left panel: two possible measurement configurations are common source-CS (top) and common drain-CD (bottom). The heater (DUT = device under test) operates in saturation while the sensor operates in the sub-threshold region. Right panel: mask image indicating both FETs being located in a common active area, spaced only one poly pitch away [36, 37].

source or common drain configuration are considered. One device functions as a *heater* while the other functions as a *sensor*. The sub-threshold slope at the sensor side varies in direct response to the temperature at the heater device side. A sample of biasing conditions together with the schematics for both configurations is given in figure 2.25 (left panel). Figure 2.25 (right panel) shows the mask image that indicates that both FETs that are located in a common active area and are separated by only one poly pitch.

The experimental procedure to estimate the hot-spot temperature was proposed by the IMEC group [38]. First, as schematically illustrated in figure 2.26, the increase in temperature (ΔT) induced by an nFET (i.e. the 'heater' from figure 2.24) is extracted by making use of an nFET sensor that is located nearby this nFET or the device under test (DUT). This is done by using *temperature dependent characteristics* of the sensor. As can be seen, the sensor is connected to a common source configuration with the heater (figure 2.25, top left panel and right panel). This configuration allows for the closest possible 'in silicon' sensor since the two devices are separated only by one gate pitch. Also, both these devices share the same active area which is surrounded by shallow trench isolation (figure 2.25, right panel). Subsequently, the sub-threshold swing (SS) in the sensor is extracted using a *modified EKV model*, as illustrated in figure 2.26 [39].

The results from the experimental data are used as a reference for the thermal simulations. The coupled solver uses the Monte Carlo method and the energy balance and Poisson's equation to simulate the heating at the device level. The Joule heating, defined as the product of the current density and the electric field, is extracted from these device-level simulations and used in the circuit level simulation. The Joule heating term is used as an input for the interconnect level solver (Giga3D module within Silvaco Atlas framework). This Silvaco model provides the temperature boundary conditions at the device-level within the global electro-thermal device simulator. To integrate and interface these two separate modules, MATLAB is used (shown in figure 2.24).

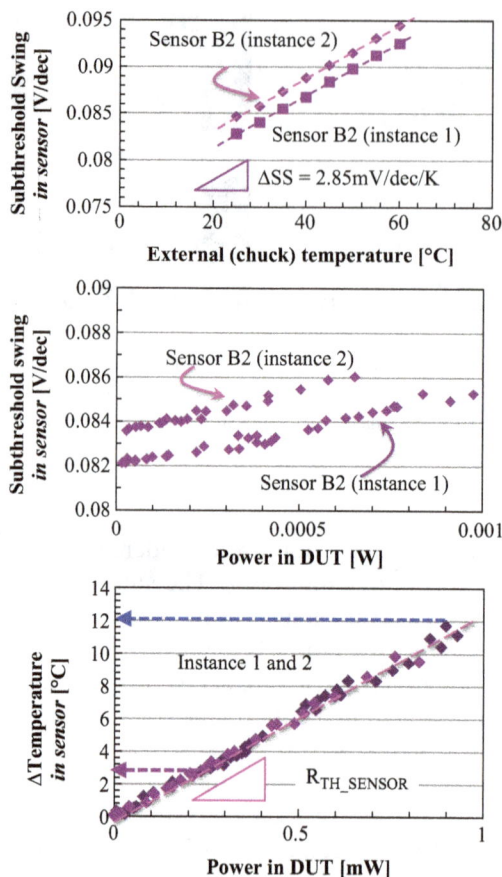

Figure 2.26. Results for two instances of a device from figure 2.24. Top panel: the subthreshold swing (SS) varies linearly with the externally applied chuck temperature. Middle panel: when drawing a large current through the heater, the SS of the sensor varies linearly. Bottom panel: using the initial SS as a reference, the extracted ΔT in the sensor gives consistent results for both instances [36, 37].

The experimental transfer characteristic curves for drain voltages $V_{DS} = 1$ V, 1.5 V and 2 V at the sensor are shown in figure 2.27 (left panel). From the extracted sub-threshold slopes, using the EKV method [40] the corresponding average sensor temperatures for different bias conditions (VDS and VGS) are shown in figure 2.27 (right panel). The goal of the multi-scale simulations is two-fold [41]: (1) Given the structure and the bias conditions in the heater-sensor configuration, the sensor temperature variation from figure 2.27 is reproduced. (2) Once the sensor temperature match is achieved, extrapolation of the peak heater (DUT) temperature is performed. Hence the temperature of the hot-spot is uncovered in an indirect way.

To account for the heating at the source, gate and drain interconnects as shown in figure 2.28, first the complete circuit (device + interconnects) domain is solved using the Giga3D module of Silvaco Atlas. The temperature at the interconnects as well as

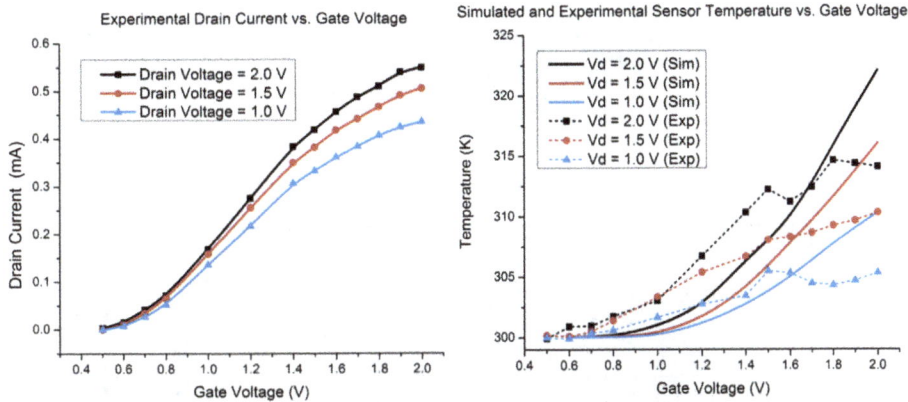

Figure 2.27. Left panel: measured transfer characteristics of the sensor. Parameter is the drain voltage $V_{DS} = 1$, 1.5 and 2 V. Right panel: extrapolated and simulated average sensor temperature for different combinations of drain and gate voltage. The EKV method is used for the extrapolation of the sensor temperature from the sub-threshold characteristics from figure 2.26.

the temperature in the device (along one cross-section) are shown in figure 2.28. At the boundaries of the rectangular cross-section from the bottom panel in figure 2.28, lattice temperature is registered and used in the thermal particle-based device simulator which then provides: (1) the actual temperature of the hot-spot, and (2) the Joule heating terms that are used back in the Giga3D module for the next Gummel iteration (outer loop) of the model. It should be noted that solving the energy balance equations for acoustic and optical phonons self-consistently with the Boltzmann transport equation for the electrons (that is solved using the MC method) is, by itself, a multi-scale problem [42]. Hence, in the implemented scheme we have three levels of abstraction. The convergence of the global Gummel loop is shown in figure 2.29.

The global loop converges within 5–10 Gummel cycles. In figures 2.30 and 2.31 the lattice (acoustic phonon) temperature and the optical phonon temperatures for the heater-sensor combination in the common-drain configuration (figure 2.24 bottom left panel) are shown. Depending upon the applied bias, the lattice temperature profile obtained from Silvaco Atlas leads to an underestimation of the hotspot temperature by about 10–20 K. Figure 2.30 shows the bottleneck in the energy transport due to the low group velocity of the optical phonons. One can see a more localized hot-spot as compared to the acoustic (lattice) temperature case [43].

2.4 Conclusions

A self-consistently coupled thermal/ensemble Monte Carlo device simulator has been described and applied to the study of fully-depleted SOI devices. We show that the pronounced velocity overshoot present in the nanometer scale device structure considered in the current study minimizes the degradation of the device characteristics due to lattice heating. This observation was also justified with Silvaco Atlas

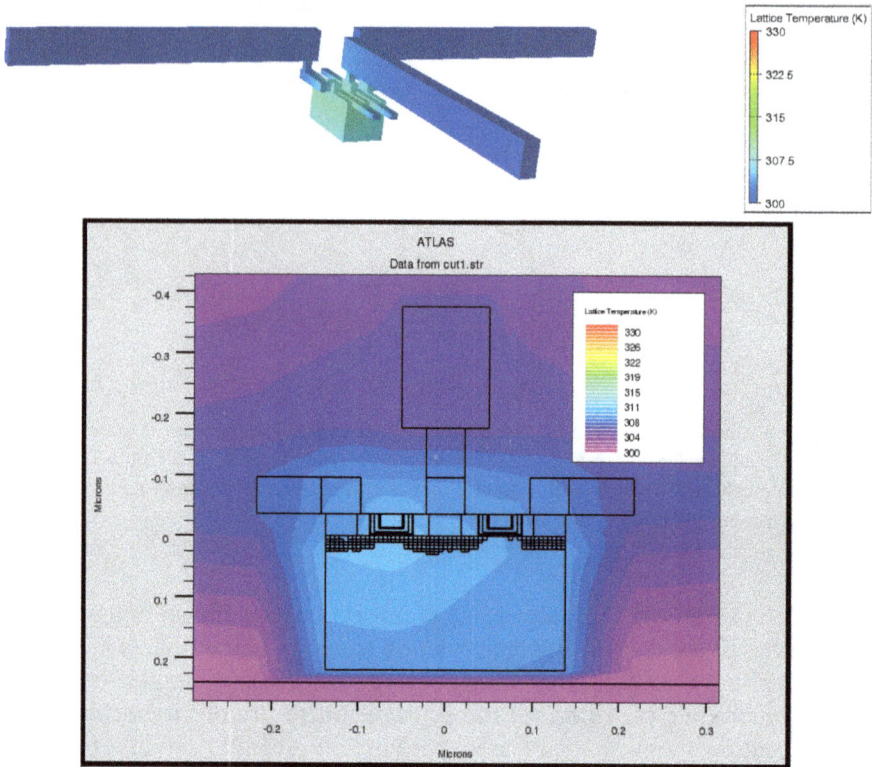

Figure 2.28. Top panel: GIGA3D modeling of the device + interconnects. Bottom panel: extracted lattice temperature boundary conditions from GIGA3D simulations. Bias conditions for the common drain configuration are: V_S (DUT) = 0, V_D (common) = 1.5 V, V_G (DUT) = 1.6 V, V_G (sensor) = 1.75 V and V_S (sensor) = 1.45 V, as shown in figure 2.25 (bottom panel).

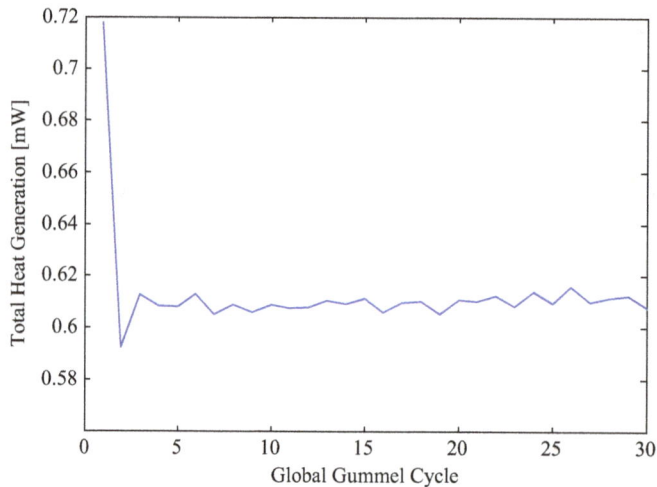

Figure 2.29. Convergence of the global Gummel cycle (loop).

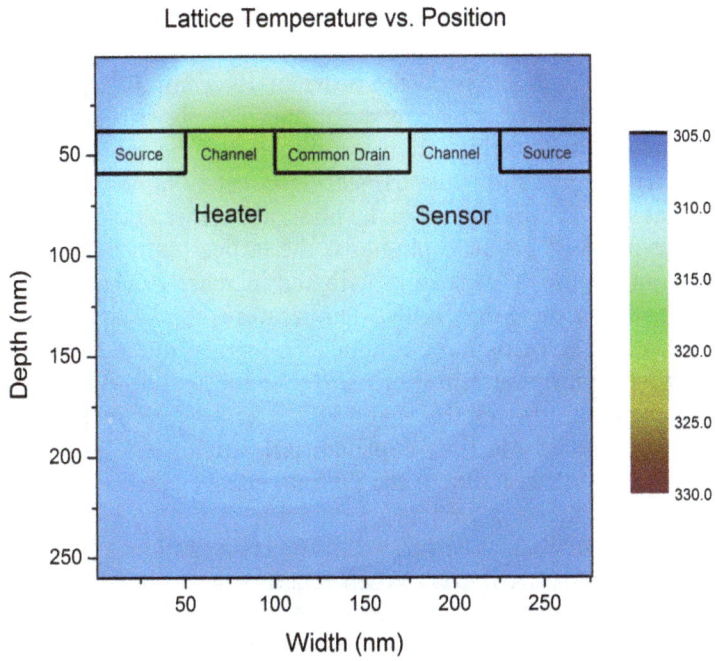

Figure 2.30. Lattice temperatures in the heater (DUT)—sensor configuration.

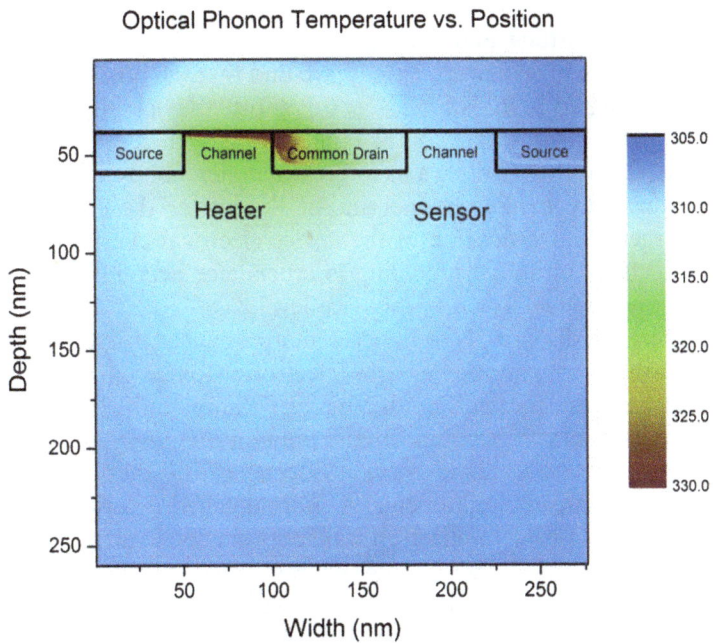

Figure 2.31. Optical phonon temperatures in the heater (DUT)—sensor configuration.

simulations that demonstrated that for larger energy relaxation times, that correspond to the case of more pronounced velocity overshoot, current degradation in the on-state due to thermal effects is on the order of 10%, not 30% as found in larger device structures in which velocity overshoot does not play a significant role.

We also presented an investigation of the influence of the gate temperature on the amount of current degradation due to heating effects. Namely, we used the gate contact as a heat sink to properly solve the phonon balance equations. As seen from the temperature maps of acoustic phonons, the lattice temperature in the source, channel and drain region is increasing with the increase of the gate temperature, which means that the increased lattice temperature has a larger impact on the decrease of the carrier velocity in the channel. To further investigate the influence of the gate temperature on the current decrease due to the heating effects, instead of the gate oxide, we used a gate stack (SiO_2 and HfO_2) and copper metal gate with finite thickness for the same 25 nm fully-depleted SOI structure. When the gate is not modeled, the current decrease due to the thermal effects is 5% more when it is used as a gate stack than when it is used as a gate oxide. Also, the lattice temperature is increased due to the gate stack compared to the corresponding lattice temperature profile when using silicon dioxide. When examining heating in different device technologies, we observed a bottleneck between the lattice and the optical phonon temperature in the channel, which is more pronounced for shorter devices, due to the fact that the energy transfer between optical and acoustic phonons is relatively slow compared to the electron-optical phonon processes and the fact that the electrons are in the velocity overshoot (and since the channel is very short, they spent little time in the channel). To better understand the phonon temperature bottleneck, different cross-sections of the lattice and the optical phonon temperature profiles in the channel direction were investigated. Briefly, we find that the bottleneck is decreasing from Si/SiO$_2$ interface to Si/BOX interface. For shorter devices, it exists in the whole channel region, which is not the case for longer devices (thicker Si-layer and longer channel length). From the results we have presented here, one can conclude that the higher the temperature in the channel and/or the longer the electrons are in the channel, the larger the degradation of the device electrical characteristics is due to self-heating effects. Multi-scale modeling is crucial for accounting of self-heating from the interconnect level down to a device level.

However, there are drawbacks to the present approaches. First, we use analytical isotropic dispersion for phonons in our electron Monte Carlo scattering rates calculation. Second, the thermal conductivity of the material is a parameter that includes a simplified way to account for the phonon boundary scattering. We also assume that there are no ballistic phonon effects, so the thermal conductivity is presumably a good transport parameter. A more detailed description of the multi-scale electron–phonon problem (electron transport occurs on a much smaller length scale when compared to the heat transport characteristic lengthscale) should involve solution of the phonon Boltzmann transport equation as well (see figure 2.32). One can also include in the loop the LAMMPS simulation package to calculate the dispersion relations for the phonon Monte Carlo or the thermal conductivity for the energy balance model. In chapter 3 of this book we describe the solution of the

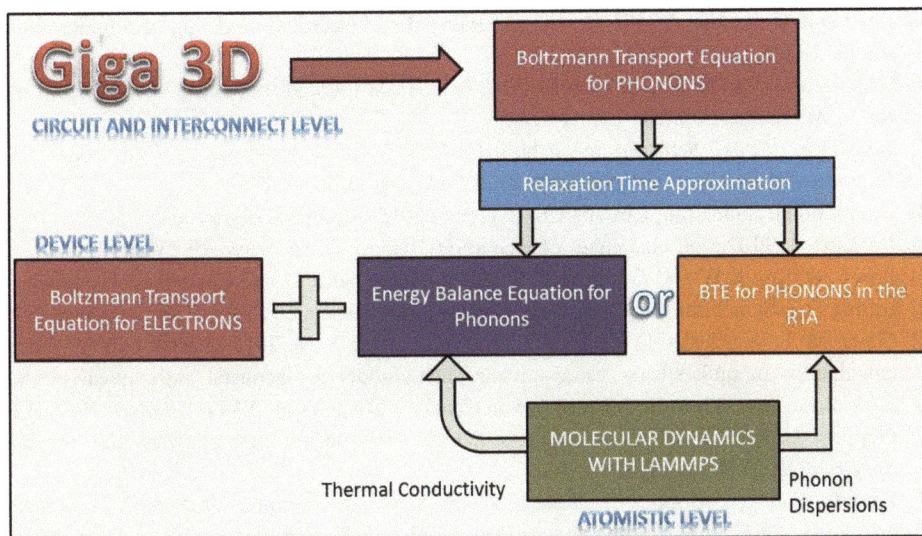

Figure 2.32. Hierarchy of models that can be applied for the solution of the multi-scale electron–phonon problem for the purpose of describing self-heating effects in ultra-nanoscale devices.

phonon Boltzmann transport equation in the relaxation time approximation (RTA) using the analytical model for the phonon dispersion relations. Full-band solution of the phonon Boltzmann Transport equation is for now a formidable task and will be addressed in the future.

References

[1] Raman A, Walker D G and Fisher T S 2003 Simulation of non-equilibrium thermal effects in power LDMOS transistors *Solid St. Electron* **47** 1265–73

[2] Pop E, Sinha S and Goodson K E 2006 Heat generation and transport in nanometer-scale transistors *Proc. IEEEE* **94** 1587–601

[3] Lai J and Majumdar A 1996 Concurrent thermal and electrical modeling of submicrometer silicon devices *J. Appl. Phys.* **79** 7353

[4] Majumdar A, Fushinobu K and Hijikata K 1995 Effect of gate voltage on hot electron and hot phonon interaction and transport in a submicrometer transistor *J. Appl. Phys.* **77** 6686

[5] Wachutka G K 1990 Rigorous thermodynamic treatment of heat generation and conduction in semiconductor device modeling *IEEE Trans. Comp. Aided Design* **11** 1141–9

[6] Gaur S P and Navon D H 1976 Two-dimensional carrier flow in a transistor structure under non-isothermal conditions *IEEE Trans. Electron Devices* **23** 50–7

[7] Leung Y K, Paul A K, Goodson K E, Plummer J D and Wong S S 1997 Heating mechanisms of LDMOS and LIGBT in ultrathin SOI *IEEE Electron Device Lett.* **18** 414

[8] Silvaco ATLAS manual http://www.silvaco.org

[9] Sadi T, Kelsall R W and Pilgrim N J 2007 Electrothermal Monte Carlo simulation of submicrometer Si/SiGe MODFETs *IEEE Trans. Electron Devices* **54** 332–39

[10] Narumanchi S V J, Murthy J Y and Amon C H 2004 Submicron heat transport model in silicon accounting for phonon dispersion and polarization *Trans. ASME* **126** 946–55

[11] Lai J and Majumdar A 1996 Concurrent thermal and electrical modeling of submicrometer silicon devices *J. Appl. Phys.* **79** 7353–61

[12] Sinha S and Goodson K E 2005 Review: Multiscale thermal modelling in nanoelectronics *Int. J. Multiscale Comput. Eng.* **3** 107–33

[13] Jang D *et al* 2015 Self-heating on bulk FinFET from 14 nm down to 7 nm node *IEEE International Electron Devices Meeting (IEDM)* pp 11.6.1–11.6.4

[14] International Solid-state Circuits Conf. Trends 2015 http://isscc.org/trends/

[15] Borkar S 1999 Design challenges of technology scaling *IEEE Micro* **19** 23–29

[16] Pop E, Dutton R W and Goodson K E 2004 Analytic band Monte Carlo model for electron transport in Si including acoustic and optical phonon dispersion *J. Appl. Phys.* **96** 4998–5005

[17] Gada M L, Vasileska D, Raleva K and Goodnick S M 2012 Electron drift velocity calculations in bulk silicon using an analytical model for acoustic and optical phonon dispersions *Tech. Proc. 2012 NSTI Nanotechnology Conf. Expo, NSTI-Nanotech 2012* 712-5

[18] Geppert L 1999 Solid state [semiconductors. 1999 technology analysis and forecast] *Spectrum IEEE* **36** pp 52–6

[19] Zeng G, Fan X, LaBounty C, Croke E, Zhang Y, Christofferson J, Vashaee D, Shakouri A and Bowers J E 2003 Cooling Power Density of SiGe/Si Superlattice Micro Refrigerators *Materials Research Society Fall Meeting 2003, Proceedings* **793** paper S2.2 Boston, MA

[20] Majumdar A 1993 Microscale heat conduction in dielectric thin films *J. Heat Transfer* **115** 71–16

[21] Chen G 2001 Ballistic-diffusive heat-conduction equations *Phys. Rev. Lett.* **86** 2297–300

[22] Enz C C and Vittoz E A 2006 *Charge-Based MOS Transistor Modeling: The EKV Model for Low-Power and RF IC Design* (New York: Wiley)

[23] Reif F 1985 *Fundamentals of Statistical and Thermal Physics* (London: McGraw-Hill)

[24] Asheghi M, Touzelbaev M N, Goodson K E, Leung Y K and Wong S S 1998 Temperature dependent thermal conductivity of single-crystal silicon layers in soi substrates *ASME J. Heat Transfer* **120** 30–33

[25] Choi S-H and Maruyama S 2003 Evaluation of the phonon mean free path in thin films by using classical molecular dynamics *J. Korean Phys. Soc.* **43** 747–53

[26] Ju Y S and Goodson K E 1999 Phonon scattering in silicon films with thickness of order 100 nm *Appl. Phys. Lett.* **74** 3005–7

[27] Liu W and Asheghi M 2004 Phonon-boundary scattering in ultra-thin single-crystal silicon layers *Appl. Phys. Lett.* **84** 3819–21

[28] Liu W and Asheghi M 2005 Thermal conductivity of ultra-thin single crystal silicon layers *J. Heat Transfer* **128** 75–83

[29] Raleva K, Vasileska D and Goodnick S M 2008 Is SOD technology the solution to heating problems in SOI devices? *IEEE Electron Device Lett.* **29** 621–4

[30] Ruxandra M, Costescu M, Wall A and Cahill D G 2003 Thermal conductance of epitaxial interfaces *Phys. Rev.* B **67** 054302

[31] For 2003 International Technology Roadmap for Semiconductors (ITRS), see website http://public.itrs.net/

[32] Chen G and Shakouri A 2002 Heat transfer in nanostructures for solid-state energy conversion *J. Heat Transfer* **124** 242–52

[33] Vashaee D and Shakouri a 2004 Electronic and thermoelectric transport in semiconductor and metallic superlattices *PJ. Appl. Phys.* **95** 1233–45

[34] Vashaee D and Shakouri a 2004 Nonequilibrium electrons and phonons in thin film thermionic coolers *Microscale Thermophys. Eng.* **8** 91–100

[35] Ziman J 2001 *Electrons and Phonons: The Theory of Transport Phenomena in Solids* (Oxford: Oxford University Press)

[36] Bury E, Kaczer B, Roussel P J, Ritzenthaler R, Raleva K, Vasileska D and Groeseneken G 2014 Experimental validation of self-heating simulations and projections for transistors in deeply scaled nodes *Proc. IEEE, Reliability Physics Symposium, IEEE International* XT. 8.1-XT. 8.6

[37] Raleva K, Bury E, Kaczer B and Vasileska D 2014 Uncovering the temperature of the hotspot in nanoscale devices *Proc. IEEE, Computational Electronics (IWCE), 2014 International Workshop* pp 104-6

[38] Hardy R J 1970 Phonon Boltzmann equation and second sound in solids *Phys. Rev.* B 1193–206

[39] Mazumder S and Majumder A 2001 Monte carlo study of phonon transport in solid thin films including dispersion and polarization *J. Heat Transfer* **123** 749–59

[40] Joshi A A and Majumdar A 1993 Transient ballistic and diffusive phonon heat transport in thin films *J. Appl. Phys.* **74** 31–39

[41] Chen G 2003 Ballistic-diffusive equations for transient heat conduction from nano to macroscales *J. Heat Transfer* **124** 320–8

[42] Chen G 1996 Non-local and non-equilibrium heat conduction in the vicinity of nanoparticles *ASME J. Heat Transfer* **118** 539–45

[43] Stillinger F H and Weber T A 1985 Computer simulation of local order in condensed phases of silicon *Phys. Rev.* B **31** 5265–71

Chapter 3

Phonon Monte Carlo simulation

As described in chapter 2, and repeated here for completeness, a phonon is a quanta of vibrational energy that arises from oscillating atoms within a crystal. Any solid crystal consists of atoms bound into a specific repeating three-dimensional spatial pattern called a lattice. The atoms are coupled together through spring-like forces, and, therefore, can oscillate about their equilibrium lattice positions, due to thermal motion or in response to outside forces. Since the atoms are coupled, this vibrational motion generates mechanical waves that carry heat and sound through the material. A packet of these waves can travel throughout the crystal with a definite energy and momentum, so in quantum mechanical terms the waves can be treated as a particle, called a phonon, where a phonon is a discrete unit or quantum of vibrational mechanical energy, just as a photon is a quantum of electromagnetic energy. Phonons and electrons are the two main types of elementary particles or excitations in solids. Whereas electrons are responsible for the electrical properties of materials, phonons determine the speed of sound within a material and how much heat it takes to change its temperature.

The transport of phonons can be studied through the solution of the Boltzmann transport equation (BTE) for phonons [1], that is of the form

$$\frac{\partial f}{\partial t} + v_g \cdot \nabla f = \frac{\partial f}{\partial t}\bigg|_{\text{scattering}}. \tag{3.1}$$

In equation (3.1), f is the distribution function of an ensemble of phonons, and v_g is the group velocity. In chapter 2 we denoted by g the phonon distribution function. Here we reverse the notation and use f for the phonon distribution function since we are entirely concerned with phonons in this chapter. The left-hand side (lhs) of equation (3.1) represents the change of the distribution function due to motion, whereas the right-hand side (rhs) represents the change in the distribution function due to scattering. The motion of phonons causes the distribution function to deviate from equilibrium while scattering tends to restore equilibrium. Although the motion of phonons is due to population gradient, it is popular to call this motion drifting of phonons instead of diffusion of phonons.

A popular way of solving the BTE is through the relaxation time approximation (RTA), where the scattering term on the rhs of equation (3.1) is approximated as

$$\left.\frac{\partial f}{\partial t}\right|_{\text{scattering}} = \frac{f - f_{eq}}{\tau}. \tag{3.2}$$

Here f_{eq} is equilibrium distribution under no forces and no gradients and τ is the total relaxation time. The RTA implies that the system decays back to equilibrium in an exponential form with a time constant equal to the total relaxation time. The total relaxation time depends on the scattering rates of different scattering mechanisms involved in the transport process. For heat conduction in crystals the scattering mechanisms are impurity scattering (or mass difference scattering includ-ing isotopes), phonon boundary scattering and phonon–phonon scattering. The total scattering rate is then calculated through Mathiessen's rule as

$$\tau^{-1} = \tau_{IM}^{-1} + \tau_{B}^{-1} + \tau_{Ph-Ph}^{-1} \tag{3.3}$$

where τ_{IM}^{-1} is the impurity scattering rate, $\tau_{B^{-1}}$ is the boundary scattering rate and τ_{Ph-Ph}^{-1} is the phonon–phonon scattering rate. The RTA is only valid when the deviation from equilibrium is small. Also note that RTA is valid only when τ is independent of the distribution function and the population gradient.

Based on the momentum conservation of the phonons, the phonon scattering mechanisms are classified into normal (N) processes and Umklapp (U) processes [1]. If the momentum is conserved (see equation (3.4)) in the scattering process, then the process is called N-type process. If the momentum is conserved (equation (3.5)) within a non-zero multiple of one of the unit reciprocal lattice vectors, $\vec{G}$, of the crystal, then the process is called U process. The total energy is conserved in both processes. Thus, for the N and U processes, we have:

$$\vec{k_1} + \vec{k_2} = \vec{k_3} \quad \text{(N Process)} \tag{3.4}$$

$$\vec{k_1} + \vec{k_2} = \vec{k_3} + \vec{G} \quad \text{(U Process)}. \tag{3.5}$$

It is well known that the N-type processes tend to shift the phonon distribution function towards high momentum in k-space, while the U processes tend towards low momentum. When equilibrium is established between absorption and emission of phonons for the two processes, it results in Bose–Einstein equilibrium distribution function for phonons. As the temperature increases, the number of phonons under-going N processes and U processes increases and a new equilibrium is established. When a temperature gradient is present, then N processes (which favor higher momentum) in higher temperature regions cause phonons to diffuse to lower temperature regions while U processes (which favors lower momentum states) resist this diffusion, which leads to finite thermal conductivity in the materials. Hence, the U process is a resistive process for thermal conductivity. Impurity and boundary scattering are also resistive processes for the thermal conductivity. For calculating the thermal conductivity, only phonon–phonon scattering is considered in this chapter.

3.1 Phonon–phonon scattering

Phonon–phonon scattering is the dominant scattering process involved in the heat conduction of materials. This scattering occurs due to the anharmonic lattice forces between atoms in the crystal (harmonic forces are responsible for the phonon description itself). Lattice forces between atoms are complicated functions of the separation distance between atoms and can be expanded using Taylor series, i.e.

$$H = H_0 + \lambda H_3 + \lambda^2 H_4 + \lambda^3 H_5 + \cdots, \qquad (3.6)$$

where H_0 is the harmonic Hamiltonian, H_3, H_4, H_5, ... are the perturbing terms involving three, four, five, ..., interacting phonons. Within the quadratic approximation (H_0), the phonons do not interact with each other and have, therefore, infinite lifetime. Higher order terms of the Taylor series expansion give rise to anharmonic effects such as scattering. Including only the cubic terms and applying first order perturbation theory results in three-phonon interaction processes. With quartic terms, four-phonon interactions are possible, etc. Developing expressions for scattering rates starting from the perturbing Hamiltonian is difficult. Figure 3.1 illustrates the possible three-phonon processes in terms of Feynman diagrams involving creation and annihilation of phonons. W1 and W4 represent the decay process of a phonon of wavevectors q' and q, respectively into two different phonons. Such processes are representative of the important decay path for phonons in semiconductors in which an excited optical mode decays into two acoustic branch modes. The reverse process is illustrated in W2 and W3.

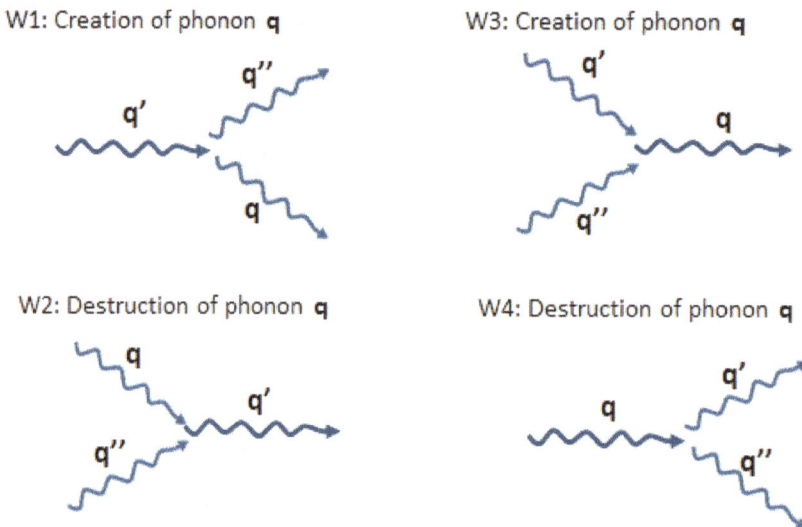

Figure 3.1. Three-phonon interaction processes that may destroy or create the phonon q.

The frequency and the temperature dependence of the three-phonon scattering rates are a strong function of the actual phonon branch and of the dispersion in the phonon spectrum. The approximate expressions typically used to describe three-phonon scattering are only valid for certain types of phonons or for a limited temperature range. Furthermore, the scattering processes are not necessarily independent, and thus simple addition of scattering probabilities may not be justifiable. This has led to an experimental determination of analytical expressions for three-phonon scattering rates. Such expressions for the relaxation times of scattering for longitudinal acoustic (LA) and transverse acoustic (TA) phonons are derived by Holland [2], and are of the form:

$$\tau_{N,LA}^{-1} = B_{LN}\omega^2 T^3 \tag{3.7}$$

$$\tau_{U,LA}^{-1} = B_{LU}\omega^2 T^3 \tag{3.8}$$

$$\tau_{N,TA}^{-1} = B_{TN}\omega T^4 \tag{3.9}$$

$$\tau_{U,TA}^{-1} = \begin{cases} 0 & \omega<\omega_{1/2} \\ \dfrac{B_{TU}\omega^2}{\sinh\left(\dfrac{\hbar\omega}{k_B T}\right)} & \omega>\omega_{1/2} \end{cases} \tag{3.10}$$

where $\omega_{1/2}$ is the TA phonon frequency corresponding to $K/K_{\max} = 0.5$ and B_{LN}, B_{LU}, B_{TN} and B_{TU} are constants for a given material. The values of these constants for Si are listed in table 3.1.

The scattering rates for longitudinal optical (LO) and transverse optical (TO) phonons are ignored mainly because of their low group velocities. Recent studies suggested, however, that optical phonons must be considered in the steady-state thermal conductivity predictions. Using the method described by Han and Klemens [3], Narumanchi *et al* [4] derived the scattering rates of U processes both for acoustic and optical modes. The scattering rates of U processes involving only acoustic

Table 3.1. Holland constants for scattering (data from [2]).

Constant	Value	Units
B_{LN}	2.0×10^{-24}	s deg^{-3}
B_{LU}	2.0×10^{-24}	s deg^{-3}
B_{TN}	9.3×10^{-13}	deg^{-3}
B_{TU}	5.5×10^{-18}	s
$\omega_{1/2}$	2.417×10^{13}	rad s^{-1}

phonons of type given by equation (3.11) are given by equation (3.12). In equation (3.11) BZB refers to the Brillouin zone boundary.

$$LA + TA(BZB) \leftrightarrow LA; \qquad TA + TA(BZB) \leftrightarrow LA;$$
$$TA + LA(BZB) \leftrightarrow LA \qquad (3.11)$$

$$\tau_{ij}^{-1} \approx \frac{\chi\gamma^2\hbar}{3\pi\rho v_{ph}^2 v_g}\omega_i\omega_{tr}\omega_j r_c^2\left[\frac{1}{e^{\hbar\omega_{tr}/k_BT}-1} - \frac{1}{e^{\hbar\omega_j/k_BT}-1}\right]. \qquad (3.12)$$

The indices refer to phonons interactions of the type given by equation (3.11), namely, an incoming phonon (Ph_i) interacts with the translated phonon (Ph_{tr}) resulting in an outgoing phonon (Ph_j), i.e.

$$Ph_i + Ph_{tr} \leftrightarrow Ph_j \qquad (3.13)$$

In equation (3.12), χ is the degeneracy of translated mode phonon, $v_g(= |\partial\omega_j/\partial q_j|)$ is the group velocity of outgoing phonon, $v_{Ph}(= \omega_i/q_i)$ is the phase velocity of the incoming phonon, γ is the Gruneisen constant with value of 0.59 for silicon, ρ is the density of the material (here Si) and r_c is the effective radius: $r_c = \frac{\frac{2\pi}{a} - k_i}{2\sqrt{2}}$, where a is the lattice constant of Si, k_i is the wavevector of the incoming phonon. For the interactions of type given by equation (3.14) involving optical phonons (indexing is the same as in equation (3.13)), the scattering rate is given by equation (3.15), hence

$$LA/TA + LA/TA \leftrightarrow LO/TO(BZB) \qquad (3.14)$$

$$\tau_{ij}^{-1} \approx \frac{\chi\gamma^2\hbar}{3\pi\rho v_{ph}^2 v_g^3}\omega_i\omega_j\omega_{tr}^3\left[\frac{1}{e^{\hbar\omega_{tr}/k_BT}-1} - \frac{1}{e^{\hbar\omega_j/k_BT}-1}\right] \qquad (3.15)$$

where all the symbols have the same meaning as in equation (3.12) except v_g. In equation (3.15) v_g is the group velocity of the translated mode phonon i.e. $v_g = \partial\omega_{tr}/\partial q_{tr}$.

The phonon–phonon scattering processes that are important in thermal conductivity calculations are listed in table 3.2.

Assuming that the scattering processes are independent, the total scattering rates of different three-phonon processes of the type given in equation (3.11) and equation (3.14) can be added using Mathiessen's rule, and the scattering rate for U process is then given as

$$\tau_{U,i}^{-1} = \sum_j\tau_{ij}^{-1}$$

Mittal and Mazumder [1] used hybrid approach to calculate the thermal conductivities of silicon thin films. They used Holland's expressions for N-processes and equations (3.12), (3.15) and (3.16) for U-processes.

In a recent paper, Kazan et al [5] used analytical expressions similar to those of Holland except that the scattering rate constants are not calibrated from experiments

Table 3.2. Important three-phonon interactions for U-processes.

Longitudinal acoustic phonons	Transverse acoustic phonons
$LA + TA(BZB) \leftrightarrow LA$	$TA + TA(BZB) \leftrightarrow LA$
$LA + TA \leftrightarrow LO(BZB)$	$TA + LA(BZB) \leftrightarrow LA$
$LA + TA \leftrightarrow TO(BZB)$	$TA + LA \leftrightarrow LO(BZB)$
$LA + LA \leftrightarrow LO(BZB)$	$TA + LA \leftrightarrow TO(BZB)$
$LA + LA \leftrightarrow TO(BZB)$	
Longitudinal optical phonons	Transverse optical phonons
$LO \leftrightarrow LA + TA(BZB)$	$TO \leftrightarrow LA + TA(BZB)$
$LO \leftrightarrow LA + LA(BZB)$	$TO \leftrightarrow LA + LA(BZB)$
$LO \leftrightarrow TA + LA(BZB)$	$TO \leftrightarrow TA + LA(BZB)$

but determined via material constants. The scattering rates as calculated by Kazan *et al* [5] are of the form

$$\tau_{U,i}^{-1} = \sum_j \tau_{ij}^{-1} \tag{3.16}$$

$$\tau_{N,LA}^{-1} = B_{LN}\omega^2 T^3 \tag{3.17}$$

$$\tau_{U,LA}^{-1} = B_{LU}\omega^2 T e^{-\theta_{DL}/3T} \tag{3.18}$$

$$\tau_{N,TA}^{-1} = B_{TN}\omega T^4 \tag{3.19}$$

$$\tau_{U,TA}^{-1} = B_{TU}\omega^2 T e^{-\theta_{DT}/3T} \tag{3.20}$$

with

$$B_{LN} = \frac{k_B^3 \gamma_L^2}{\rho \hbar^2 v_L^3} \tag{3.21}$$

$$B_{LU} = \frac{\hbar \gamma_L^2}{M v_L^2 \theta_{DL}} \tag{3.22}$$

$$B_{TN} = \frac{k_B^4 \gamma_T^2}{\rho \hbar^3 v_T^5} \tag{3.23}$$

$$B_{TU} = \frac{\hbar \gamma_T^2}{M v_T^2 \theta_{DT}} \tag{3.24}$$

Table 3.3. Material parameters for calculating scattering constants (data taken from [6]).

Name	Value	Units
v_T	5840	m s^{-1}
v_L	8430	m s^{-1}
θ_{DT}	240	K
θ_{DL}	586	K

Table 3.4. Material parameters for calculating scattering constants (data taken from [7]).

Name	Value	Units
γ_L	1.1	no units
γ_T	0.6	no units
B_{TU}	1.0×10^{-19}	s^{-1}K^{-3}
B_{LU}	5.5×10^{-20}	s^{-1}K^{-3}
B_{TN}	7.1×10^{-13}	s^{-1}K^{-5}
B_{LN}	2.4×10^{-24}	s^{-1}K^{-5}

where v_L, v_T are group velocity of LA, TA phonons at the Brillouin Zone center, θ_{DL}, θ_{DT} are the Debye temperatures for LA and TA phonons, γ_L, γ_T are Gruneisen constants for LA and TA phonons, respectively. The values taken from the literature [6] are listed in table 3.3.

The values for Gruneisen constants and scattering rate constants, taken from [7] are summarized in table 3.4.

Three ways of calculating the thermal conductivity are presented in this book. Within the first method, Holland's expressions are used and thermal conductivities of silicon are calculated without including optical phonons. In the second approach, Mittal's and Mazumder's hybrid method is used to calculate thermal conductivities of silicon with and without the inclusion of optical phonons. The third method utilizes the expressions of Kazan *et al* to calculate thermal conductivities of silicon without optical phonons. All the results are then compared with the experimentally measured thermal conductivities of silicon measured by Glassbrenner and Slack [7].

3.2 Monte Carlo simulation procedure

In this section, the solution of the phonon BTE in the relaxation-time approximation, using the Monte Carlo method, is presented. A direct method for solving the phonon BTE exists in principle, but treating different scattering mechanisms independently is very difficult, if not impossible. Also, incorporation of new scattering mechanisms is complicated. For those interested, the formalism for direct solution of the phonon BTE can be found in [4]. The main advantages of using the Monte Carlo technique are:

- Simple treatment of the transient problem.
- The ability to consider complex geometries.
- The possibility to examine each scattering process independently.

The Monte Carlo method provides good insight about the physical nature of the transport problem. New scattering mechanisms can be easily incorporated into the simulation code without much difficulty. The main steps involved in Monte Carlo solution method for the phonon BTE are:

- Input geometry and boundary conditions.
- Input material constants and dispersion relation.
- Calculate scattering, polarization, distribution and weighted distribution tables.
- Initialize phonons in the simulation domain (cells) with distribution tables.
- Free-flight the phonons.
- Calculate the new temperature ($\tilde{T}$).
- Scatter phonons with scattering rates at $\tilde{T}$.
- Using weighted distribution tables, add or remove phonons (reinitialization).
- Calculate the phonon flux flow in all cells.
- Thermalize the end cells with the temperature boundary conditions.

3.2.1 Geometry and boundary conditions

Before the simulation begins, the geometry of the simulation domain has to be decided. Once the geometry is decided, the simulation domain is partitioned into cells. The cells touching the boundaries are termed thermal boundary cells. The boundary conditions set the temperature of the boundary cells. Other cells, which are not boundary cells have adiabatic boundary conditions, and hence are termed as adiabatic boundary cells. The phonons which reach the thermal boundary cells are thermalized to the boundary temperature. The phonons in the adiabatic boundary cells reaching the side boundaries are specularly reflected. If randomizing boundary scattering is incorporated, this condition is changed to partially specular and partially diffusive reflection. In the present description, boundary scattering is not considered.

The geometry chosen in the current study is a simple cubic cell stack as shown in figure 3.2. The current Monte Carlo method solves the phonon transport in

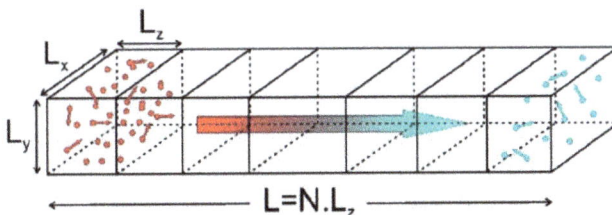

Figure 3.2. Geometry of the structure chosen for simulation in the present study. (Courtesy of [8].)

1 dimension (along the Z-direction) in silicon slab with dimensions $L_x = 50$ nm, $L_y = 50$ nm, $L_z = 400$ nm and $N = 20$. The boundary conditions are given by $T_h = T + 10K$ and $T_c = T - 10K$ where T is the temperature at which the thermal conductivity calculation is performed.

3.2.2 Material constants and phonon dispersion relations

All the material constants, including the scattering rate constants, are input to the simulator. The material chosen in this work is silicon. The dispersion of the material should be known for each phonon branch. The full band dispersion of silicon is not incorporated in the present work. Assuming isotropic Brillouin zone (BZ) approximation, the experimental phonon dispersion for silicon is fit with quadratic curve fit as

$$\omega_k = \omega_0 + v_s k + ck^2. \tag{3.25}$$

For acoustic modes, the constants v_s and c are chosen so as to capture the slope of the dispersion curve near the center of the BZ and the maximum at the edge of the BZ. For longitudinal optical phonons, they ensure the LO frequency at the BZ equals the maximum frequency of the LA mode. For both TA and TO phonons, the curves are fit such that the slopes at the BZ edge are zero. The curve fit parameters for silicon are shown in table 3.5 [9]. The phonon dispersion of silicon with the curve fitting is as shown in figure 2.13 in chapter 2 and repeated here for completeness in figure 3.3.

From the dispersion relation group velocities for the phonons are calculated using

$$v_g = \frac{\partial \omega}{\partial k} = v_s + 2ck. \tag{3.26}$$

In the present theoretical model, since an isotropic BZ is assumed, the direction of the group velocity will be the same as the direction of the wavevector k.

3.2.3 Scattering, polarization and distribution tables

In previous studies of phonon Monte Carlo simulations [10, 11], the scattering rates, probability of scattering, probability for polarization and probability for phonon modes are all calculated on the fly during the simulation. The disadvantage of this method is that adding new scattering mechanism and including optical modes

Table 3.5. Curve-fit parameters for silicon dispersion data. (Data taken from [9].)

Phonon type	ω_0 $(10^{13}$ rad s$^{-1})$	v_s $(10^3$ m s$^{-1})$	c $(10^{-7}$ m^2 s$^{-1})$
LA	0.00	9.01	−2.01
TA	0.00	5.23	−2.26
LO	9.88	0.00	−1.60
TO	10.20	−2.57	1.12

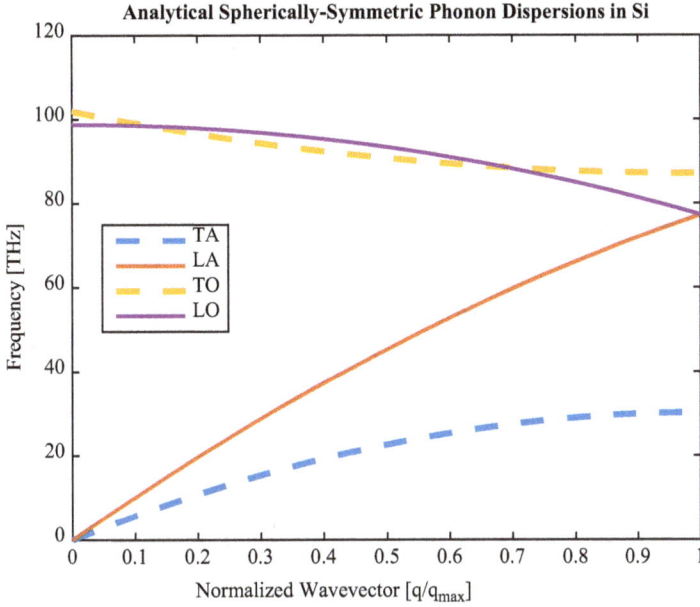

Figure 3.3. Dispersion of silicon with curve fitting. (Courtesy of [9].)

becomes difficult. A better way of simulating is to calculate tables for scattering rates, scattering probabilities, polarization probabilities and distribution probabilities. All the tables are functions of phonon frequencies and lattice temperatures. Hence, the frequency of phonons from 0 to ω_{max} is divided into 1000 bins and lattice temperature from $T_h = T + 10$ K to $T_c = T - 10$ K into 1000 bins. Let i denote the index for phonon frequency bin and j denote the index for lattice temperature bin. First, the spectral distribution of phonons at each temperature bin for each phonon mode is calculated and stored. The number of phonons of a given type t (acoustic or optical) with a given polarization p (longitudinal or transverse) in a volume of V within the spectral bin ω to $\omega + \Delta\omega$ at a given temperature T, with isotropic approximation is given by

$$N_{t,p} = \left[\frac{1}{e^{\hbar\omega_{t,p}/k_B T}-1}\right]\frac{k_{t,p}^2}{2\pi^2 v_{g;t,p}}\chi_{t,p}\Delta\omega. \tag{3.27}$$

Here $k_{t,p}$ is the wavevector of the phonon, $\chi_{t,p}$ is the degeneracy of the phonon (because of the isotropic assumption), and $v_{g;t,p}$ is the group velocity of phonon. Hence,

$$N_{\text{type}}(i, j) = \left[\frac{1}{e^{\hbar\omega_{\text{type}}/k_B T_j}-1}\right]\frac{k_{\text{type}}^2}{2\pi^2 v_{g;\text{type}}}\chi_{\text{type}}\Delta\omega \tag{3.28}$$

for type $= LA, TA, LO, TO.$

The spectral distribution of phonon is needed to build the cumulative number distribution tables. Next, the cumulative number distribution table is calculated using

$$F(i, j) = \frac{\sum_{ii=1}^{i} N_{\text{tot}}(ii, j)}{\sum_{ii=1}^{N_b} N_{\text{tot}}(ii, j)}, \tag{3.29}$$

where N_b is the total number of frequency bins (1000) and $N_{\text{tot}}(i, j)$ is given as

$$N_{\text{tot}}(i, j) = N_{LA}(i, j) + N_{TA}(i, j) + N_{LO}(i, j) + N_{TO}(i, j). \tag{3.30}$$

This cumulative number distribution table is used during the initialization of phonons to decide the frequency for phonon so that they follow Bose–Einstein statistics. The use of this table is discussed in the initialization section. An energy table is also calculated using

$$E(j) = \sum_{i=1}^{N_b} N_{\text{tot}}(i. j) \hbar \omega_i V_{\text{cell}}, \tag{3.31}$$

where V_{cell} is the volume of the cell. The polarization probability table is calculated as

$$P(i, j, 1) = \frac{N_{LA}(i, j)}{N_{\text{tot}}(i, j)} \tag{3.32}$$

$$P(i, j, 2) = \frac{N_{LA}(i, j) + N_{TA}(i, j)}{N_{\text{tot}}(i, j)} \tag{3.33}$$

$$P(i, j, 3) = \frac{N_{LA}(i, j) + N_{TA}(i, j) + N_{LO}(i, j)}{N_{\text{tot}}(i, j)}. \tag{3.34}$$

The polarization table is useful during the initialization process to determine the type and polarization of the initialized phonon. The procedure to determine the polarization of the phonon is discussed in the initialization step.

Using the scattering rates given in equations (3.7)–(3.24) above, the scattering table is calculated as

$$S_{\text{type}} = \frac{\tau_{N,\text{type}}^{-1}(i, j)}{\tau_{N,\text{type}}^{-1}(i, j) + \tau_{U,\text{type}}^{-1}(i, j)}. \tag{3.35}$$

The scattering table is used during the scattering of phonons to determine the type of scattering process a phonon would undergo in a scattering event. This is discussed in the scattering section.

Also, the probability table for being scattered within Δt of simulation time is calculated as

$$pScat_{\text{type}} = 1 - \exp\left(-\frac{\Delta t}{\tau_{N,\text{type}}^{-1}(i, j) + \tau_{U,\text{type}}^{-1}(i, j)}\right) \tag{3.36}$$

for type = LA, TA, LO, TO.

These tables are required during the scattering of phonons to decide whether a given type of phonon would undergo a scattering event. They are also used to calculate the weighted distribution and polarization tables.

The weighted cumulative distribution function is calculated as

$$F_w(i, j) = \frac{\sum_{ii=1}^{i} N_{\text{tot},w}(ii, j)}{\sum_{ii=1}^{N_b} N_{\text{tot},w}(ii, j)} \tag{3.37}$$

where $N_{\text{tot},w}(i, j)$ is given by

$$N_{\text{tot},w}(i, j) = \sum_{\text{type}=LA,TA,LO,TO} N_{\text{type}}(i, j) pScat_{\text{type}}(i, j). \tag{3.38}$$

The weighted probability distribution function is calculated as

$$P_w(i, j, 1) = \frac{N_{LA}(i, j) pScat_{LA}(i, j)}{N_{\text{tot},w}(i, j)} \tag{3.39}$$

$$P_w(i, j, 2) = \frac{N_{LA}(i, j) pScat_{LA}(i, j) + N_{TA}(i, j) pScat_{TA}(i, j)}{N_{\text{tot},w}(i, j)} \tag{3.40}$$

$$
P_w(i, j, 3)
$$
$$
= \frac{N_{LA}(i, j) pScat_{LA}(i, j) + N_{TA}(i, j) pScat_{TA}(i, j) + N_{LO}(i, j) pScat_{LO}(i, j)}{N_{\text{tot},w}(i, j)}. \tag{3.41}
$$

These tables are used in the re-initialization of phonons. The need for these tables is discussed in the scattering section and how they are used is discussed in re-initialization section.

3.2.4 Initialization

After building scattering, distribution, polarization and weighted distribution tables, the phonons in each cell have to be initialized with position vector, wave vector, frequency, group velocity and type of phonon. To do so, the total number of phonons in each cell should be known. The total number of phonons in a cell at T can be calculated as

$$N_{\text{tot}}(j) = \sum_{i=1}^{N_b} N_{\text{tot}}(i, j) \times V_{\text{cell}} \tag{3.42}$$

where V_{cell} is the volume of cell, j is the index of the bin within which T lies i.e. T lies between bin j and bin $j + 1$.

The total number of phonons in each cell is very large (for example the number of phonons in a cell of dimension 50 nm × 50 nm × 200 nm slab of silicon at 300 K is about 31 million). Hence a weighting factor is used to scale down this number. After fixing the number of phonons, the following procedure is performed until the total energy within the cell matches with the calculated energy $E(j)$ to initialize each phonon:

- Generate a random number r uniformly distributed between 0 and 1.
- Find the index i such that $F(i, j) < r < F(i + 1, j)$ where j is fixed by the temperature of the cell.
- Set the frequency of the phonon as

$$\omega = \omega_i + (2r - 1)\frac{\Delta\omega}{2}.$$

- Generate another random number r between 0 and 1.
- Find the index ii such that $P(i, j, ii) < r < P(i, j, ii + 1)$
- Based on the value of ii set the type of phonon.

$$\begin{cases} \text{type} = LA & \text{if} \quad ii = 0 \\ \text{type} = TA & \text{if} \quad ii = 1 \\ \text{type} = LO & \text{if} \quad ii = 2 \\ \text{type} = TO & \text{otherwise} \end{cases}$$

- Using the dispersion relation corresponding to the type of phonon, calculate the magnitude of q vector and set the group velocity v_g for the phonon at that q.
- Generate another random number r between 0 and 1.
- Fix θ such that $\cos(\theta) = 2r - 1$.
- Generate another random number r between 0 and 1.
- Fix ϕ such that $\phi = 2\pi r$.
- Set the q vector components as

$$q_x = q \sin(\theta)\cos(\phi); \qquad q_y = q \sin(\theta)\sin(\phi); \qquad q_z = q \cos(\theta). \qquad (3.43)$$

- Initialize the position such that the phonon will be within the boundary of the cell.

The above procedure ensures that the initialized phonons follow Bose–Einstein statistics and have zero net momentum.

3.2.5 Free-flight of phonons

After initialization, free-flight and scattering of phonons is performed until the end of the simulation time. During the free-flight time, the phonons move within the cell with a corresponding group velocity v_g. The phonon position is updated after a time of Δt as

$$\vec{r}_{new} = \vec{r}_{old} + \vec{v}_g \cdot \Delta t. \qquad (3.44)$$

If the phonons cross the boundary in the X or Y direction, then they are reflected (adiabatic boundary) and the corresponding direction of the momentum is reversed. If the phonons cross the boundary in the z direction, then it is marked so as to delete it in the present cell and update it into new cell. The marked phonons are also useful to calculate the flux flowing in and out of each cell. All the cells are updated at the end of free-fight of phonons with the drifted phonons.

3.2.6 New temperature calculation ($\tilde{T}$)

After the free-flight of phonons is completed, the total energy of the cell will change because of the drift. This means the phonon distribution is also changed. Assuming the new phonon distribution is near to Bose–Einstein equilibrium distribution, a new temperature ($\tilde{T}$) is calculated such that the total energy of the cell is the same as the total energy of the equilibrium phonon distribution at $\tilde{T}$. The calculation is done by numerically inverting equation (3.31). Since the energy table gives the table of energy dependence for a given temperature, a simple search will give the value for $\tilde{T}$. Once the value of $\tilde{T}$ is known, it is used to find the scattering rate and the probability of scattering of phonons during the scattering event. This process is explained in the next section.

3.2.7 Scattering of phonons

In the present model, phonon–phonon scattering is treated as a three-phonon phenomena. When a phonon is scattered it either combines with a second phonon and gives a third phonon or decays into two phonons, as illustrated in figure 3.1. In either case, a scattering event transforms a phonon to another phonon along with addition of a new phonon or deletion of the old phonon. Addition happens when it decays to two phonons and deletion happens when it combines to give a third phonon. Also, the energy and net momentum are conserved. So every scattering event requires a selective addition or deletion of phonon from the simulation domain. A search for such selective phonons in the simulation domain is computationally very expensive. To avoid this, previous studies on phonon Monte Carlo methods provided an approximate method of simulating the scattering event.

It is well known that scattering events tend to restore equilibrium. Since scattering is performed after every increment, Δt, of time, which is chosen such that $\Delta t > 3\tau_{max}$, 95% of the system will be restored to equilibrium. This is because the relaxation time approximation shows the system will restore equilibrium in exponential decay fashion with time constant τ and hence within three time constants, 95% will be restored. Thus, the system is in near equilibrium at t and $t + \Delta t$. Mazumder and Majumdar [12] suggested that instead of adding or deleting phonons one at a time during each scattering event, one should mark all the phonons that are going to scatter. After marking they redistribute only the marked phonons such that they follow equilibrium statistics at $\tilde{T}$. Then, they add or delete phonons until the total energy is conserved. This procedure is summarized below:

1. For each phonon in the cell, find the type, frequency (i.e. bin index i) of phonon.
2. Get a random number r between 0 and 1.
3. Mark the phonon if r is greater than $pScat_{type}(i, j)$ where j refers to temperature bin index corresponding to $\tilde{T}$.
4. For each marked phonon, get a random number between 0 and 1.

5. Find the index ii such that $F(ii, j) < r < F(ii + 1, j)$.
6. Change the state of phonon from bin i to ii by resetting the momentum and group velocity (position remains the same).
7. Calculate the gain of energy as $\hbar(\omega_{ii} - \omega_i)$.
8. Calculate the net gain of energy by summing the gain of all marked phonons.
9. If the net gain is negative add phonons just like in initialization step until the net gain is zero.
10. Else, delete phonons by drawing random number between 0 to N_{cell} until the net gain is zero (N_{cell} is the number of phonons in the cell).

Later Lacroix *et al* changed the distribution function $F(i, j)$ to weighted or modulated distribution function $F_w(i, j)$ in order to have the rate of creation of phonons into a state equal to its rate of destruction. This can be understood by carefully analyzing equation (3.2) which can be rewritten as

$$\left[\frac{\partial N}{\partial t} \right]_{\text{scattering}} = \frac{N - \tilde{N}_{\text{eq}}}{\tau_{\text{tot}}} \tag{3.45}$$

$$\Delta N(t) = \Delta N(0) \cdot e^{\left(-\frac{t}{\tau_{\text{tot}}} \right)} \tag{3.46}$$

$$N_{(t+\Delta t)} - \tilde{N}_{\text{eq}} = (N_t - \tilde{N}_{\text{eq}}) \cdot e^{\left(-\frac{\Delta t}{\tau_{\text{tot}}} \right)} \tag{3.47}$$

$$N_{(t+\Delta t)} = \tilde{N}_{\text{eq}} \left(1 - e^{\left(-\frac{\Delta t}{\tau_{\text{tot}}} \right)} \right) + N_{\text{drift}} e^{\left(-\frac{\Delta t}{\tau_{\text{tot}}} \right)}. \tag{3.48}$$

Note that equation (3.47) is written for the scattering events between t and $t + \Delta t$ and in equation (3.48) N_t is nothing more than the spectral distribution obtained after the drifting of phonons at time t, which is denoted as N_{drift}. Mazumder and Majumdar [13] choose $t > 3 \times \tau_{\text{tot}}$, which then leads to $e^{\frac{-\Delta t}{\tau_{\text{tot}}}} \to 0$ and hence the distribution at $t + \Delta t$ is the equilibrium distribution at $\tilde{T}$. Lacroix's [8] argument is approximate in the sense that only the drift term exponential is set to zero and the exponential in equilibrium (which is nothing but a probability of scattering) should not be ignored. Hence, the equilibrium distribution is modulated or weighted with the probability of scattering.

In the scattering routine, only marking of phonons is done. The task of adding and deleting phonons is completed by invoking the reinitialization routine.

3.2.8 Reinitialization

Once the phonon is marked to undergo scattering, the act of adding or deleting the phonons is completed by the reinitialization routine performing steps 5 to 10 listed in the scattering section. The reinitialization is performed only for scattered phonons, but not for all the phonons in the cell. The scattering for boundary cells is not calculated as they should be thermalized to the boundary conditions, although drifting of phonons is done for the boundary cell.

3.2.9 Phonon flux calculation

During the drifting of phonons, the phonons that cross the boundary are marked. If the phonons move out from a $cell_i$ to $cell_{i+1}$, then the net flux crossing is added to $cell_i$. If the phonons are moving out from $cell_{i+1}$ to $cell_i$, then the net flux crossing is subtracted from $cell_i$. The net flux calculation is done with the following expression.

$$flux(\Phi_i) = \left(\sum_{\substack{cell_i \to cell_{i+1} \\ \text{all phonons } j}} \hbar\omega_j \frac{q_{j,z}}{|q_j|} - \sum_{\substack{cell_{i+1} \to cell_i \\ \text{all phonons } j}} \hbar\omega_j \frac{q_{j,z}}{|q_j|} \right) \Big/ (\Delta t \cdot A) \qquad (3.49)$$

where A is the area of cross section $(=L_x \times L_y)$. The calculated cell flux and temperature are written to a file and the thermal conductivity is calculated using

$$\kappa = - \left\langle \frac{\Phi_i}{\Delta T_i \cdot \Delta z} \right\rangle . \qquad (3.50)$$

3.2.10 Thermalization of boundary cells

The boundary cells are thermalized to the boundary cell temperature. All the phonons in the boundary cell are initialized similar to the normal initialization process, but at the boundary cell temperature. This process is repeated for every time step of the simulation.

3.3 Verification of Monte Carlo code

The Monte Carlo steps are implemented in Fortran language. The basic framework for the PMC is adopted from Ramayya et al [10] and has been modified to implement the scattering tables and optical phonons transport for heat conduction. First, the input material properties for the silicon are verified by plotting the dispersion curves and group velocity curves. The dispersion curves for TA, LA, TO and LO phonons calculated from this work and Mittal and Mazumder [1] were shown in figure 3.3. The corresponding group velocities are compared in figure 3.4.

Next, the cumulative distribution function (CDF) curve is verified. Since these CDF curves only depend on dispersion curves and Bose–Einstein distribution, they are independent of scattering rates and the results from different work can be compared for verification. The CDF with optical phonons at different temperatures comparing with Mittal and Mazumder [1] is shown in figure 3.5. The CDF without optical phonons at different temperatures comparing with [1] is shown in figure 3.6.

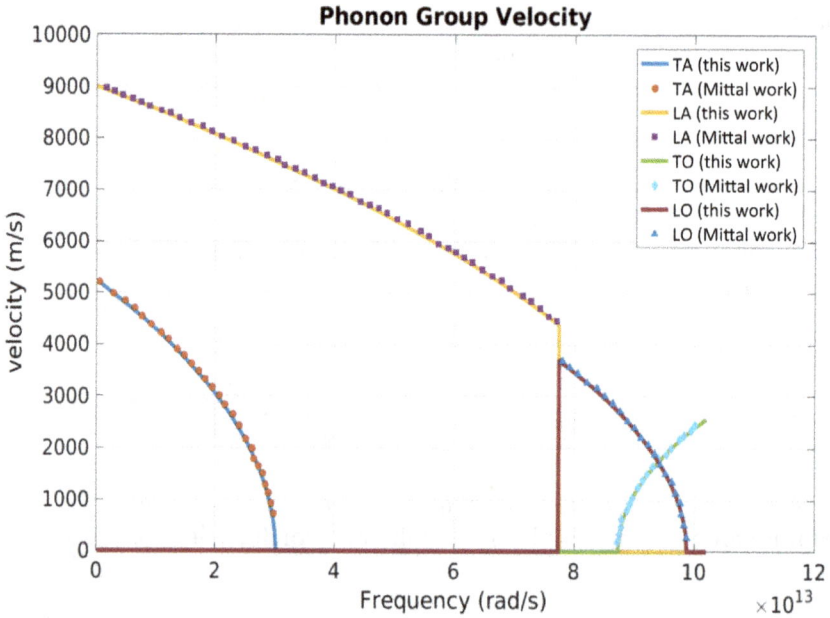

Figure 3.4. Comparison of the group velocities of silicon (data taken from A Mittal [1]).

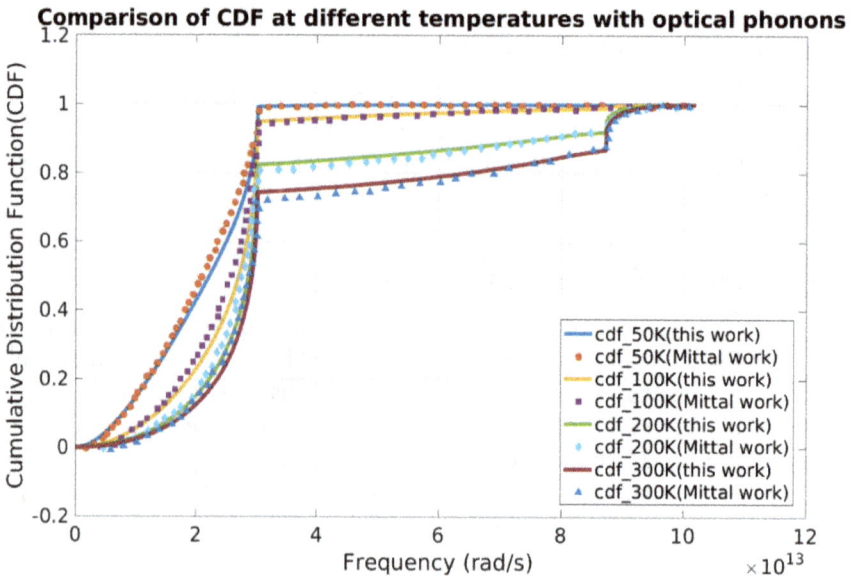

Figure 3.5. Comparison of the cumulative distribution function at different temperatures with optical phonons (data from Mittal and Mazumder [1]).

Noting that there is a slight difference in the result from [1] and the current work, another comparison for the CDF without optical phonons is performed with data from Lacroix [8]. Figures 3.7 and 3.8 show the CDF at 300 K and 500 K of the present work compared with data extracted from Lacroix [8].

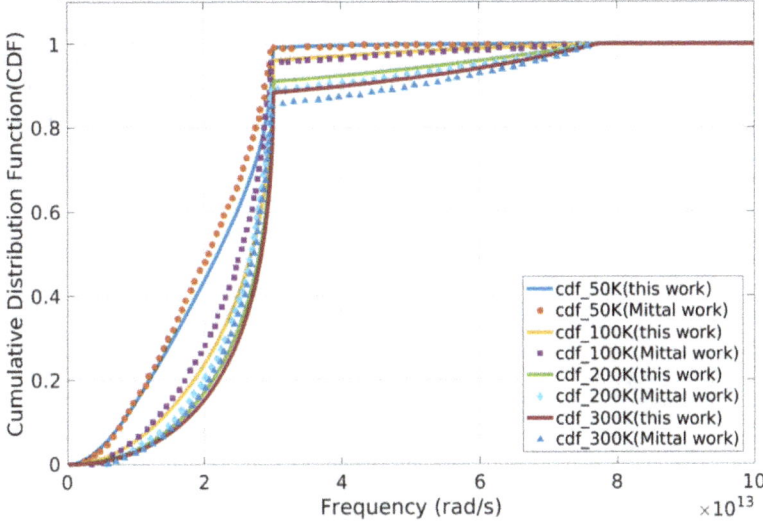

Figurer 3.6. Comparison of the cumulative distribution function at different temperatures without optical phonons (data from Mittal and Mazumder [1]).

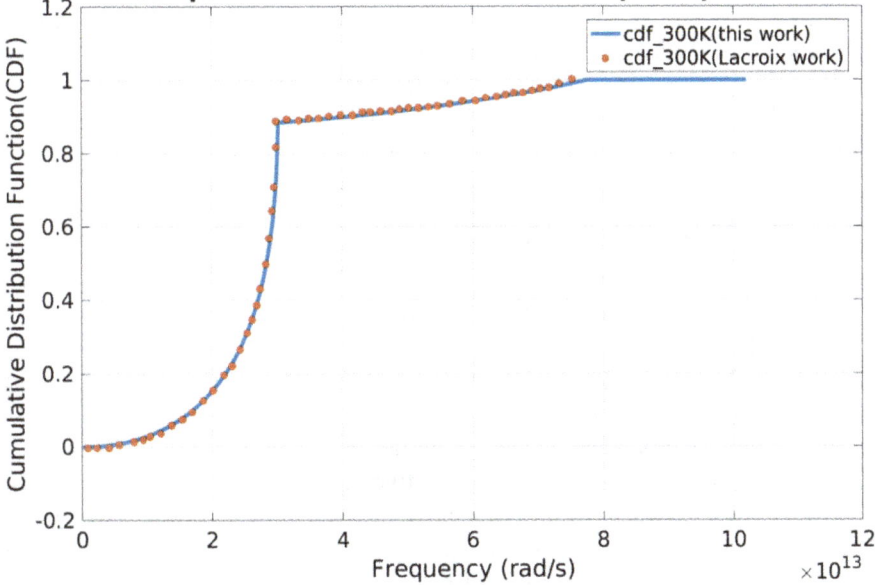

Figure 3.7. Comparison of the CDF at 300 K without optical phonons (data from Lacroix [8]).

The next set of data verified is the scattering rates. As discussed earlier, the scattering rates are calculated in three different ways. Using Holland's expression, the scattering times at 100 K and 300 K are plotted in [1]. These scattering times are compared with the present work and are shown in figure 3.9 (300 K) and figure 3.10 (100 K).

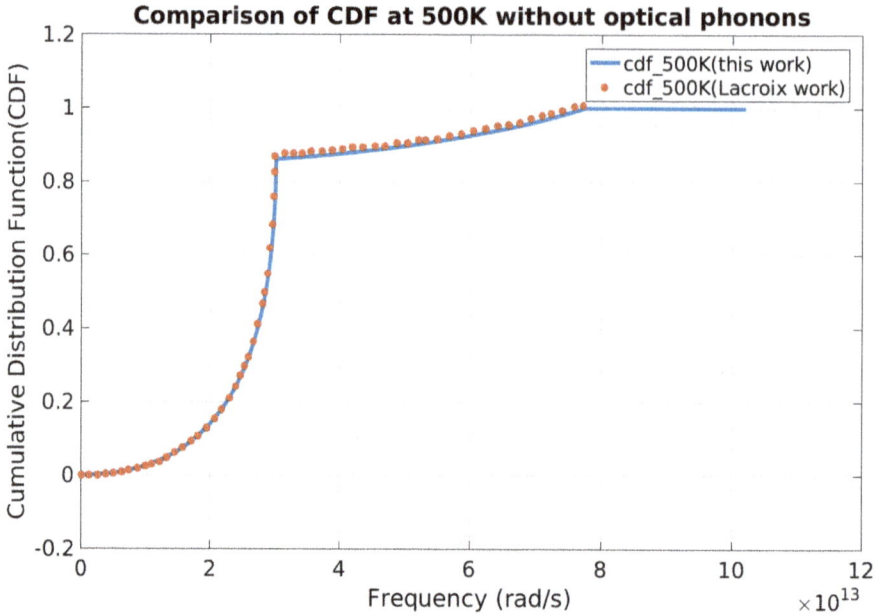

Figure 3.8. Comparison of the CDF at 500 K without optical phonons (data from Lacroix [8]).

For Monte Carlo particle based simulations, one observed behavior is the shift of the drift velocity from zero along the direction of the flow of particles. Applying a gradient of 400 K around 300 K ($T_c = 100$ K and $T_h = 500$ K) in the z-direction, the average velocity of the phonons are as shown in figure 3.11.

3.4 Phonon Monte Carlo results

In this section, the results for the scattering tables, polarization tables and distribution tables are presented. The tables depend on the temperature and frequency of the phonons. The total scattering times calculated with three different approaches at 300 K are as shown in figure 3.12. The cumulative and weighted cumulative distribution functions at 300 K is as shown in figures 3.13 and 3.14, respectively. The results for polarization tables and modulated polarization tables are as shown in figures 3.15 and 3.16, respectively. From the scattering time plots, it is clear that high frequency phonons have low scattering times, implying they participate in scattering events more frequently. From the CDF and weighted CDF tables, it is clear that the probability of being a transverse acoustic phonon is very high and hence their population is higher than the longitudinal phonons at TA phonon frequencies. This is also evident from the polarization table plots shown in figures 3.15 and 3.16.

The thermal conductivity variation versus temperature is calculated for bulk Si through different approaches. Table 3.6 lists all the thermal conductivity calculation methods. The thermal conductivity of bulk Si calculated through the methods listed

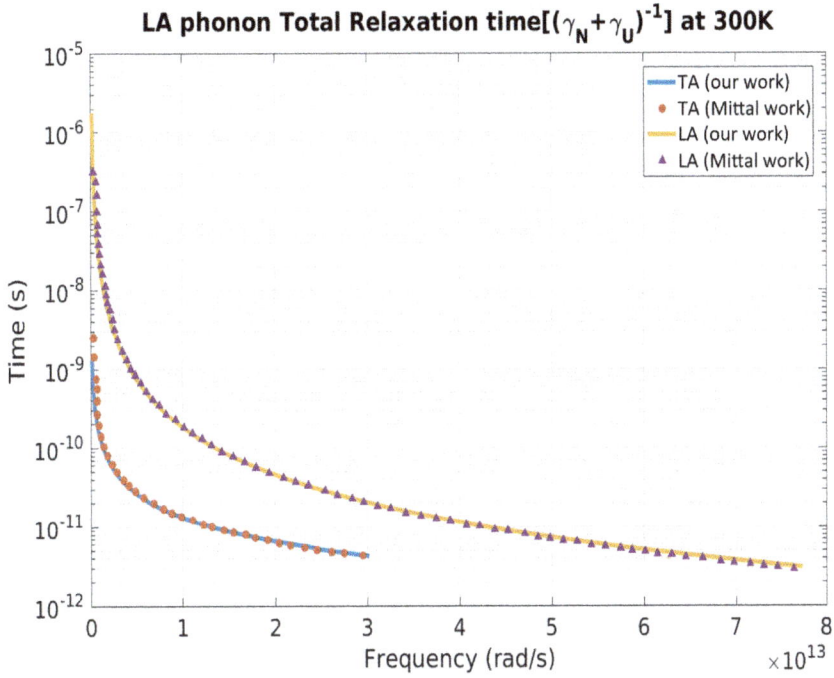

Figure 3.9. Scattering times comparison at 300 K (data from Mittal and Mazumder [1]).

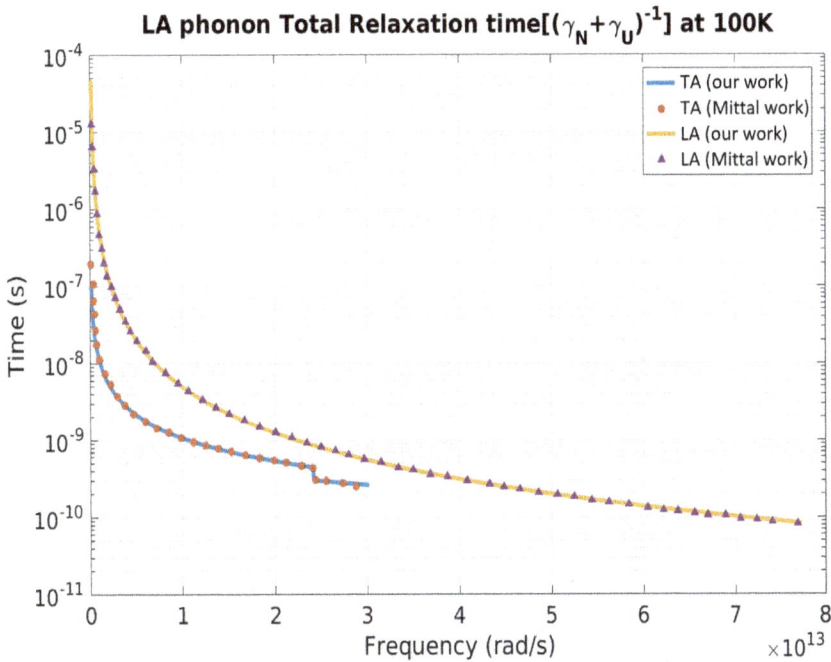

Figure 3.10. Scattering time comparison at 100 K (data from Mittal and Mazumder [1]).

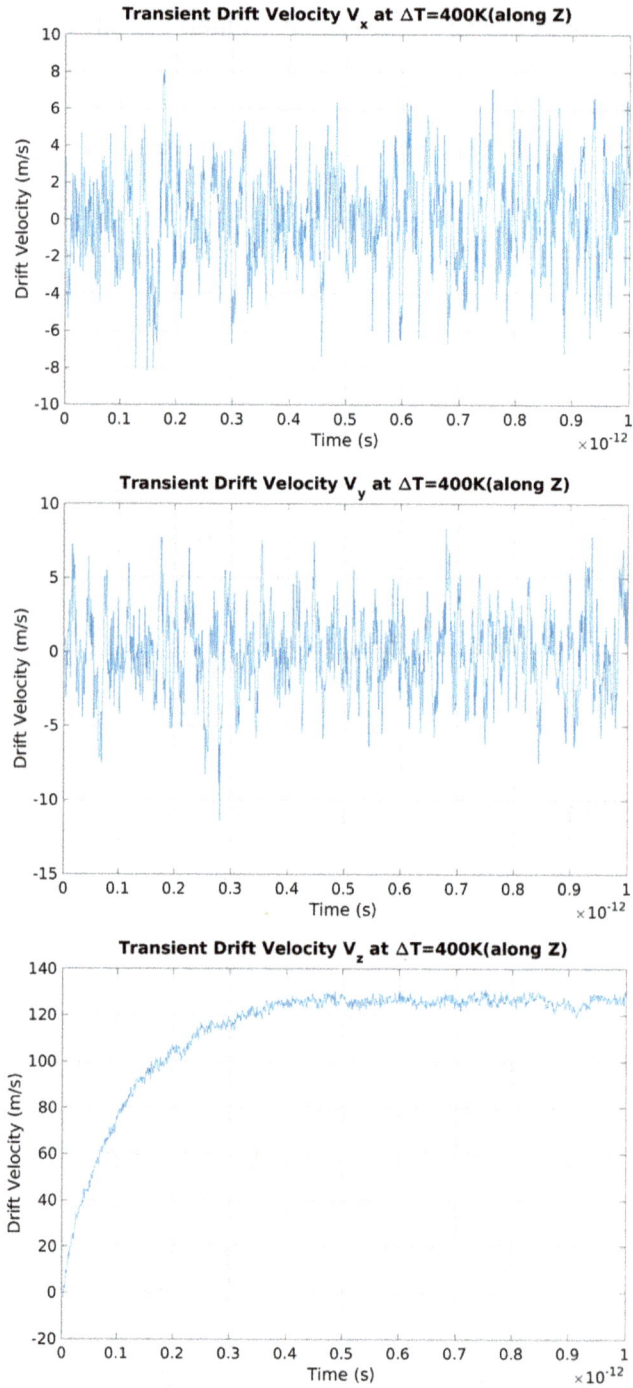

Figure 3.11. Average drift velocities of phonons in the *x, y* and *z*-direction when a temperature gradient of 400 K is present along the *z*-direction.

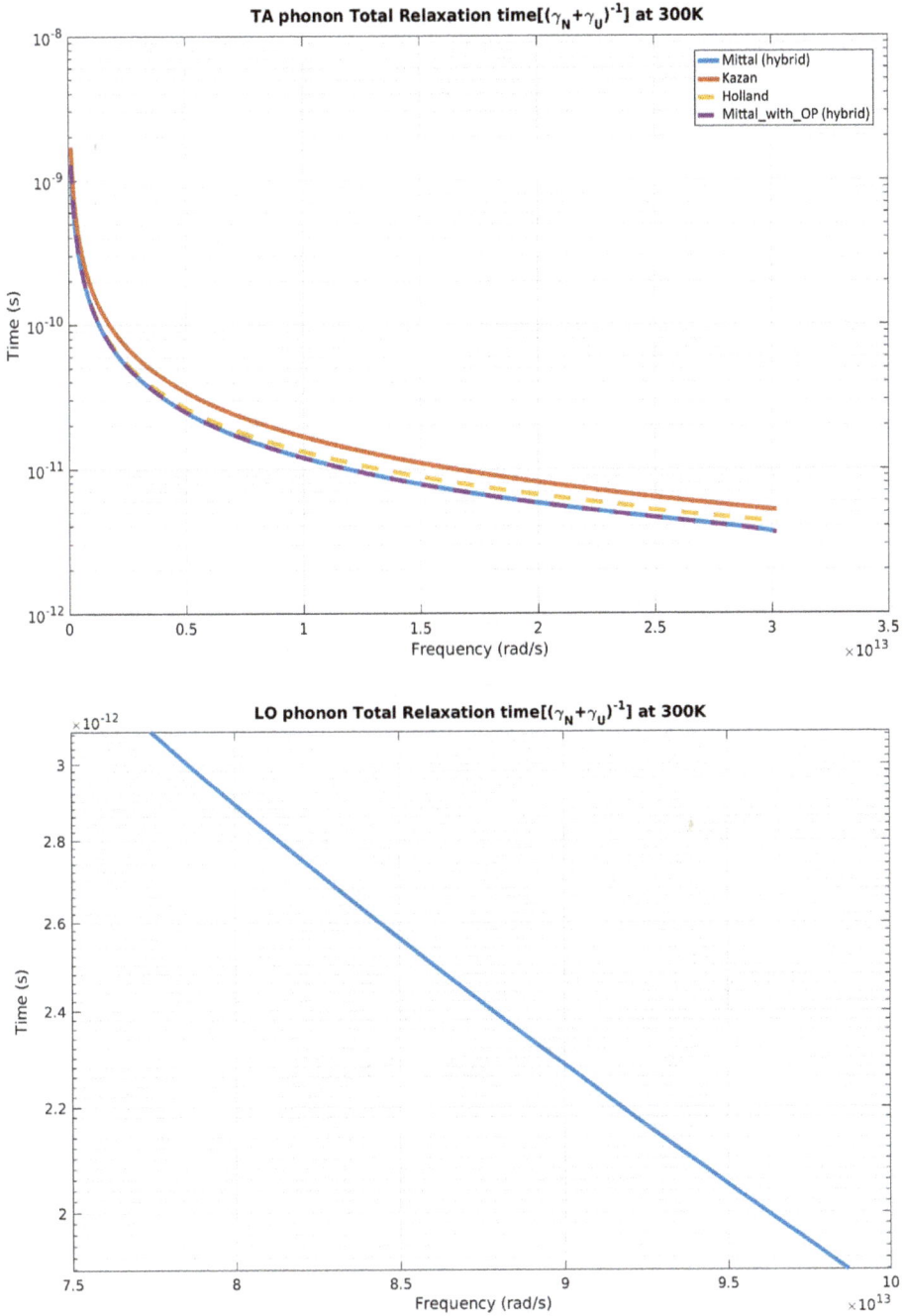

Figure 3.12. Scattering time comparison calculated from different approaches at 300 K.

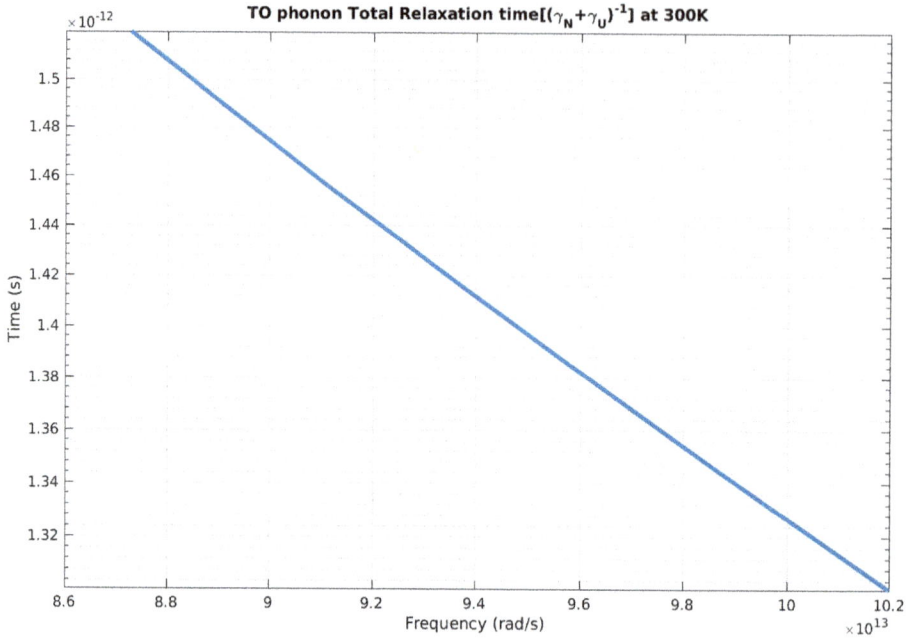

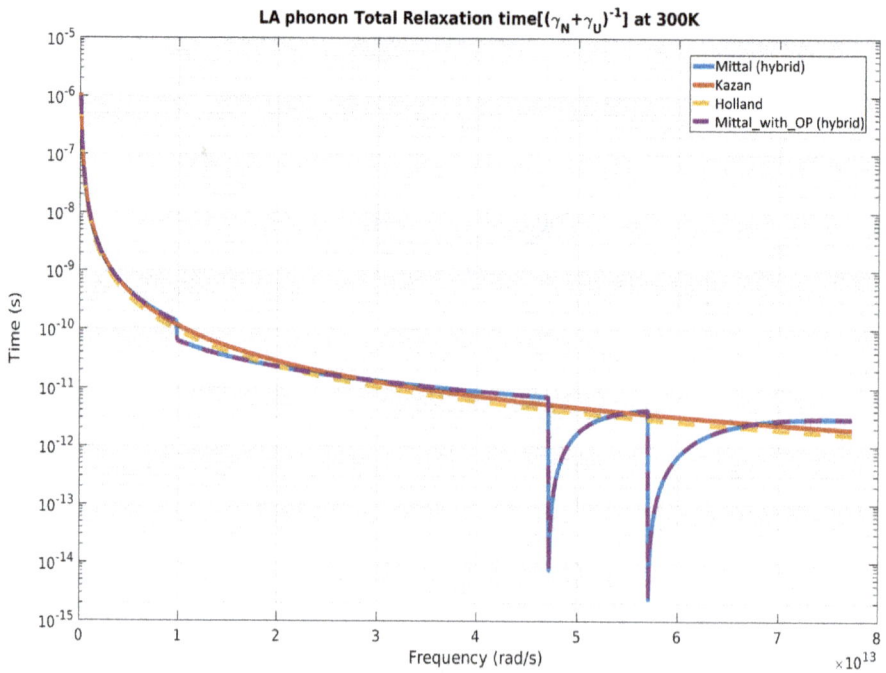

Figure 3.12. (Continued.)

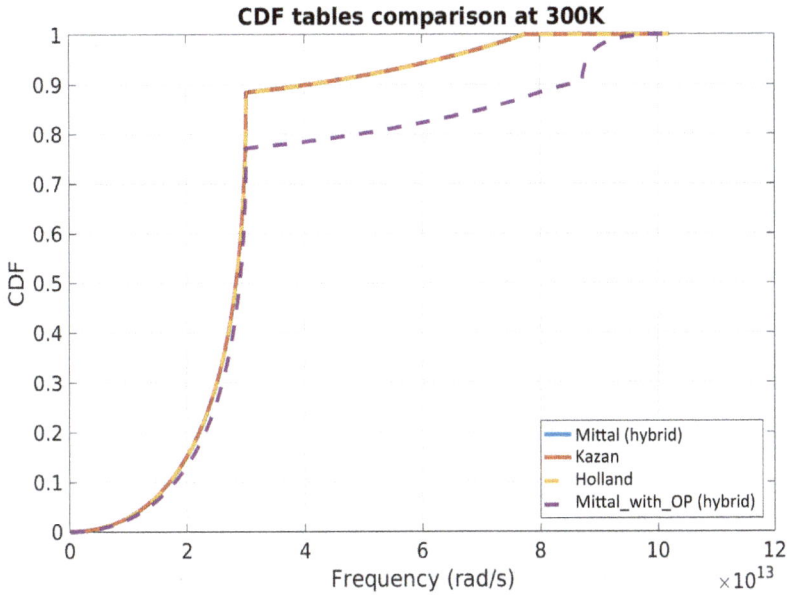

Figure 3.13. CDF comparison calculated from different approaches at 300 K.

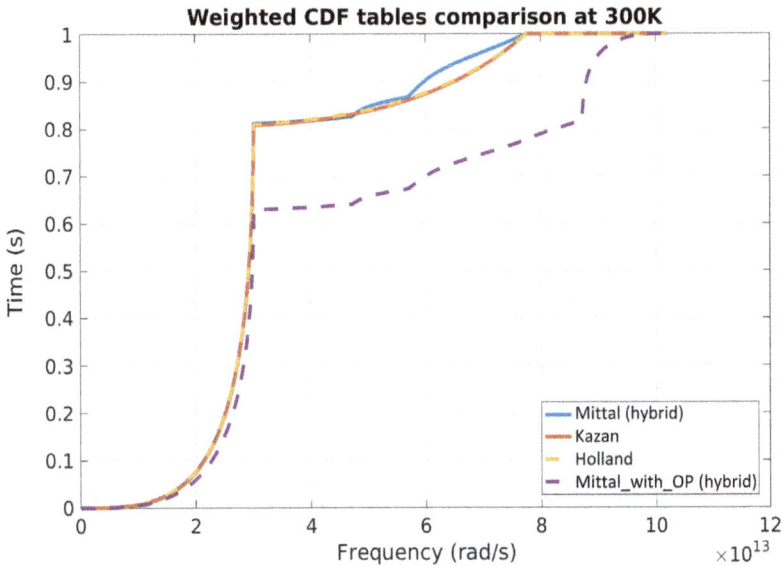

Figure 3.14. Weighted CDF comparison calculated from different approaches at 300 K.

in table 3.6, is as shown in figures 3.17, 3.18 and 3.19. From the results presented, it is evident that Mazumder's way of simulating scattering events with Holland's scattering rate and Lacroix's way of simulating scattering events with calibrated Holland scattering rates are the best fits explaining the bulk thermal conductivity dependence of temperature in Si material system.

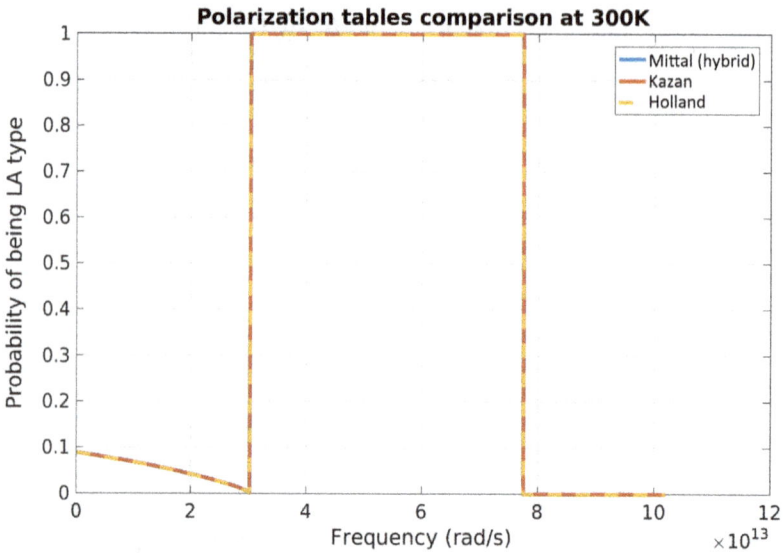

Figure 3.15. Polarization table comparison calculated from different approaches at 300 K.

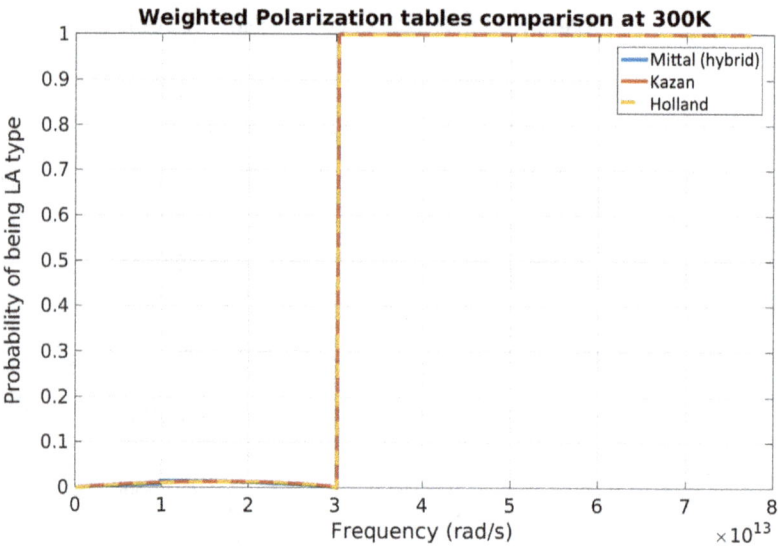

Figure 3.16. Weighted polarization table comparison calculated from different approaches at 300 K.

The temporal thermal profiles for different methods with Holland's scattering rates at 300 K are as shown in figures 3.20, 3.21 and 3.22. They clearly show that Mazumder's approach for simulating the scattering events does not provide a good representation of the thermal profile. Lacroix's method is better, but the predicted thermal conductivity is still not close to the experimental data. The calibrated

Table 3.6. Results classification based on scattering rates and scattering approaches.

Scattering rates	Scattering approach	Optical phonons	Number
Holland	Mazumder	no	1
Mittal	Mazumder	no	2
Kazan	Mazumder	no	3
Mittal	Mazumder	yes	4
Holland	Lacroix	no	5
Mittal	Lacroix	no	6
Kazan	Lacroix	no	7
Mittal	Lacroix	yes	8
Holland(calibrated)	Lacroix	no	9
Mittal(calibrated)	Lacroix	no	10
Kazan(calibrated)	Lacroix	no	11
Mittal(calibrated)	Lacroix	yes	12

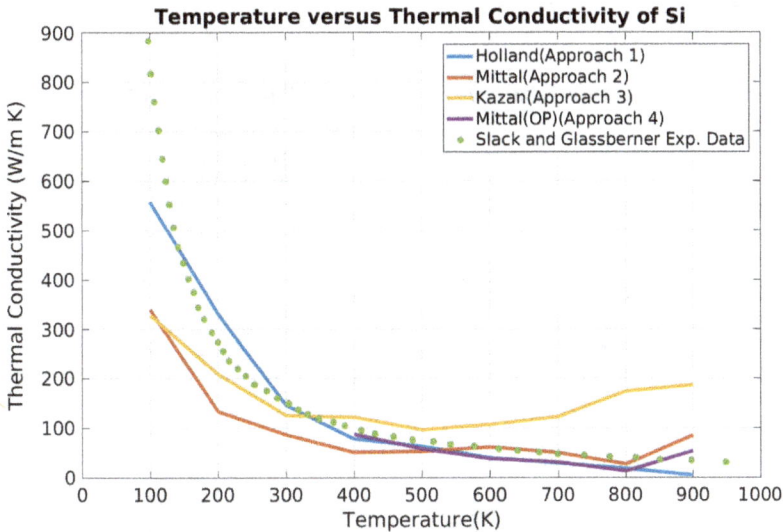

Figure 3.17. Thermal conductivity of silicon calculated through Mazumder's method.

Holland scattering rates produce a closer match with Lacroix's approach of simulating scattering events.

3.5 Conclusions

To properly study the heat transport within the device a state of the art Monte Carlo device simulator is required. In this regard a phonon Monte Carlo simulator is developed. Phonons are quasi particles that carry heat energy. Like electrons, phonons also have the Boltzmann transport equation which can be used to study their transport. Direct solution of BTE for phonons is possible but it is difficult to

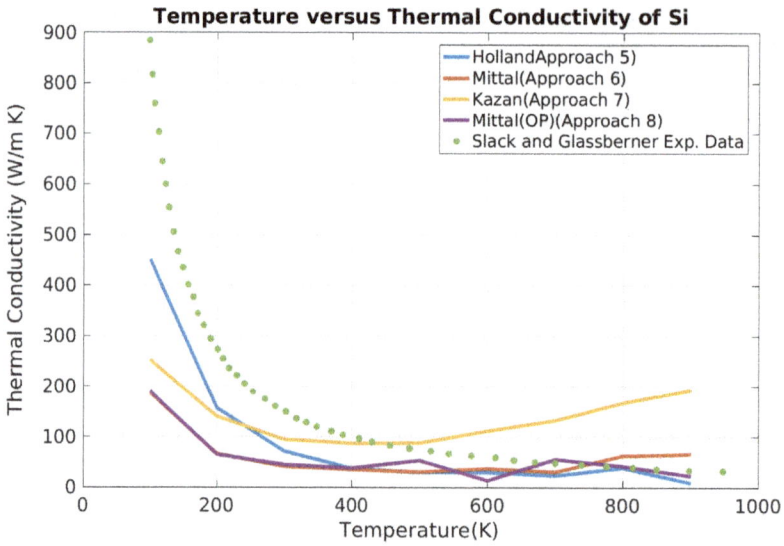

Figure 3.18. Thermal conductivity of silicon calculated through Lacroix's method.

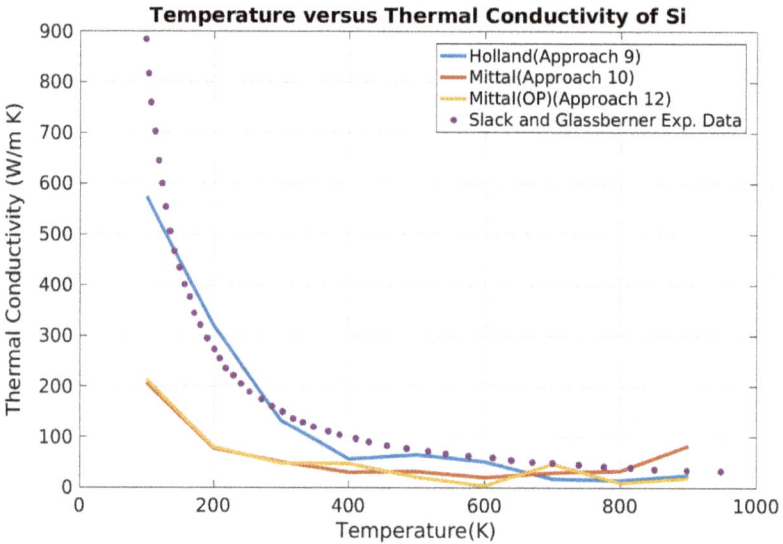

Figure 3.19. Thermal conductivity of silicon calculated through Lacroix's method with calibration of the scattering rates.

incorporate all scattering mechanisms. In the Monte Carlo based solution method it is easy to incorporate different scattering mechanisms. Although the method is computationally intensive it provides good insights into the physical nature of the transport problem. Hence the Monte Carlo based technique is used for studying phonon transport. Monte Carlo simulations require calculation of the scattering

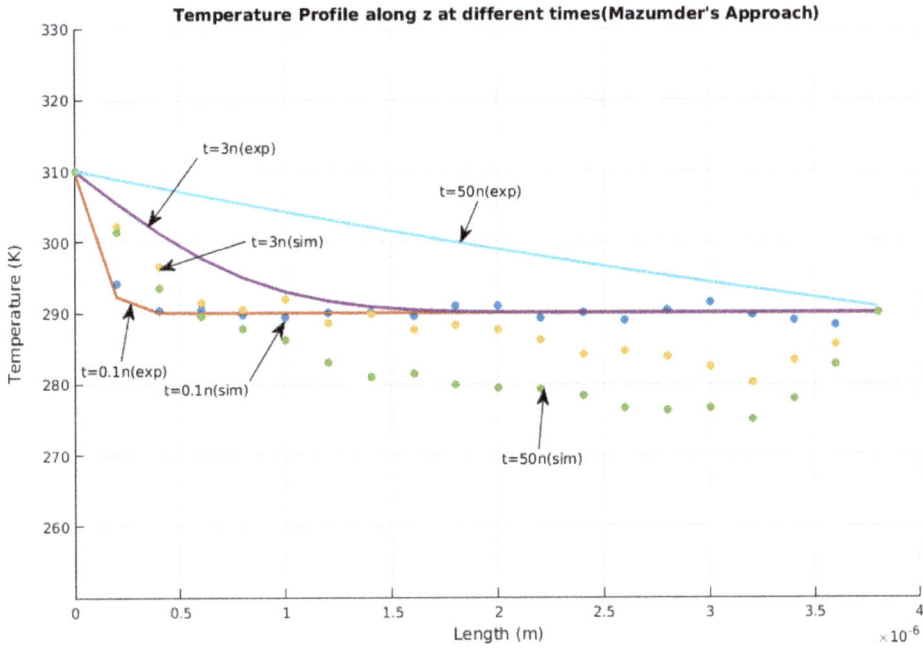

Figure 3.20. Transient behavior of the temperature profile through Mazumder's approach simulated with Holland scattering rates.

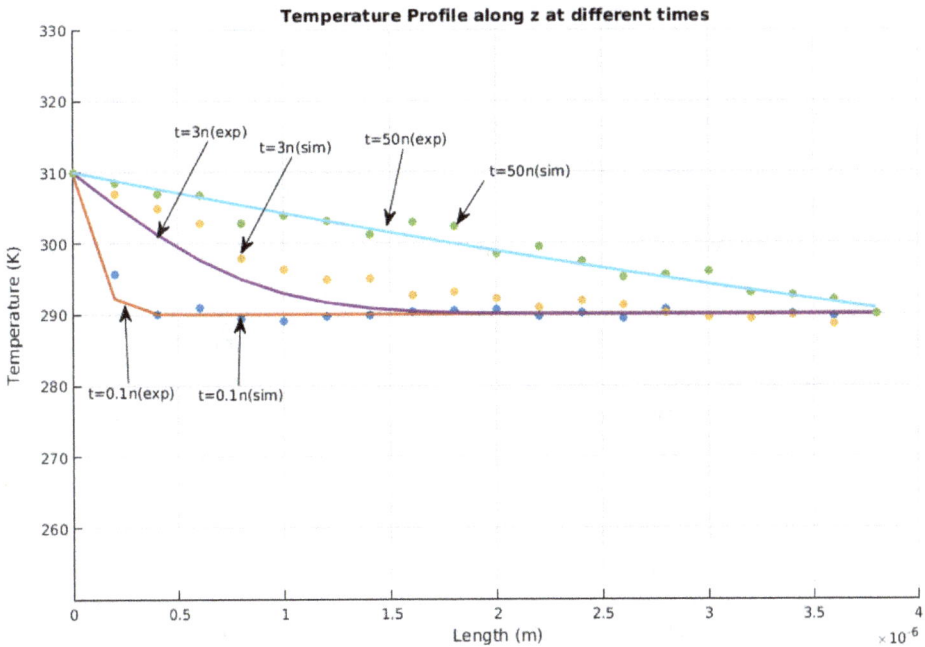

Figure 3.21. Transient behavior of the temperature profile through Mazumder's approach simulated with Holland scattering rates.

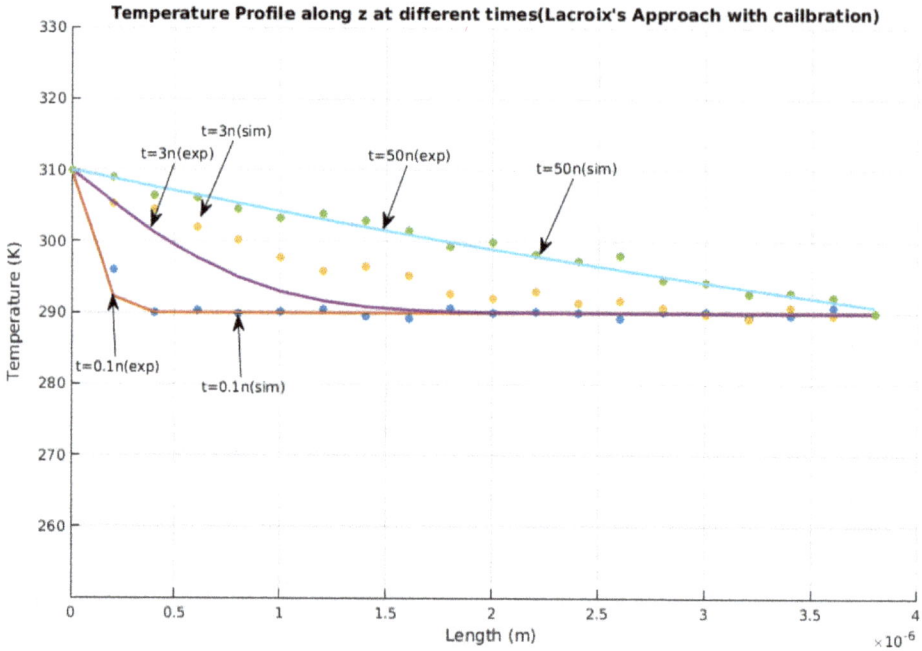

Figure 3.22. Transient behavior of the temperature profile through Lacroix's approach, simulated with calibrated Holland scattering rates.

rates for different scattering processes. In the present work scattering rates for three phonon interactions are calculated from different approaches presented in the literature. Optical phonons are also included in the transport problem. Finally, the method is applied to calculate the temperature dependence of the thermal conductivity for silicon in the range from 100 K to 900 K.

References

[1] Mittal A and Mazumder S 2010 Monte Carlo study of phonon heat conduction in silicon thin films including contributions of optical phonons *J. Heat Transfer* **132** 052402
[2] Holland M G 1963 Analysis of lattice thermal conductivity *Phys. Rev.* **132** 2461–71
[3] Han Y J and Klemens P G 1993 Anharmonic thermal resistivity of dielectric crystals at low temperature *Phys. Rev.* B **48** 6033-42
[4] Narumanchi S V J, Murthy J Y and Amon C H 2005 Submicron heat transport model in silicon accounting for phonon dispersion and polarization *J. Heat Transfer* **136** 946–55
[5] Kazan M, Guisbiers G, Pereira S, Correia M R, Marsi P, Bruyant A, Volz S and Royer P 2010 Thermal conductivity of silicon bulk and nanowires: Effects of isotopic composition, phonon confinement and surface roughness *J. Appl. Phys.* **107** 083503
[6] Morelli D, Heremans J P and Slack G A 2002 Estimation of the isotopic effect on the lattice thermal conductivity of group IV and group III–V semiconductors *Phys. Rev.* B **66** 195304
[7] Glassbrenner C J and Slack G A 1963 Thermal conductivity of silicon and germanium from 3 K to the melting point *Phys. Rev* **134** 1058–69

[8] Lacroix D, Joulain K and Lemonnier D 2005 Monte Carlo transient phonon transport in silicon and germanium at nanoscale *Phys. Rev.* B **72** 064305

[9] Pop E, Dutton R W and Goodson K E 2004 Analytic band Monte Carlo model for electron transport in Si including acoustic and optical phonon dispersion *J. Appl. Phys.* **96** 4998–5005

[10] Ramayya E B 2010 *Thermoelectric Properties of Ultrascaled Nanowires* PhD Thesis Department of Electrical Engineering, University of Wisconsin-Madison

[11] Yoo S K 2015 The phonon Monte Carlo simulation *MA Thesis* School of Electrical, Computer and Energy Engineering, Arizona State University

[12] Mazumder S and Majumdar A 2001 Monte Carlo study of phonon transport in solid thin films including dispersion and polarization *J. Heat Transfer* **123** 749–59

[13] Pop E, Banerjee K, Sverdrup P, Dutton R and Goodson K 2001 Localized heating effects and scaling of sub-0.18 micron CMOS devices *IEDM Techn. Dig.* **679**

Chapter 4

Summary

The problem of heat generation and degradation of device characteristics is an issue that is well known to the power electronics community. With applied voltages of over 40 V, carriers occupy states very high in energy and transfer it to the lattice, mainly through interactions with optical phonons [1]. Power devices are typically large in size and the drift-diffusion model for the electron and hole transport coupled with a heat conduction model, in which a Joule heating term is used, has been extensively used to model the behavior of these devices. In a few instances, the hydrodynamic model has been used for modeling of the carrier (electrons and holes) transport. For digital circuits operated at much lower voltages, self-heating (that manifests itself in a mobility degradation at high current densities) has been much less of a problem in conventional Si MOSFETs because the silicon body has very large thermal conductivity and most of the heat is dissipated through the substrate. The problem arises in silicon-on-insulator (SOI) devices (fully-depleted (FD) SOI devices, FinFETs, dual-gate (DG) structures, etc) which have already replaced conventional MOSFET devices for the 22 nm technology node [2,3] There are several problems associated with SOI technology: (1) the thermal conductivity of SiO_2 is about 100 times smaller than that of Si, so the underlying BOX acts as a barrier to heat flow (it can be considered as a large thermal resistor). The only way that heat may be transported is via the side boundaries, source and drain contacts and the gate contact. (2) Silicon thin films have a thermal conductivity smaller than bulk Si because the phonon mean free path in Si is very large, and phonon boundary scattering reduces the thermal conductivity below its bulk value [4]. The reduction of the thermal conductivity is even more pronounced in silicon nanowires because of the stronger boundary scattering [5]. Therefore, the thickness dependence of the thermal conductivity in thin silicon layers in principle has to be predicted using the phonon Boltzmann transport equation (BTE) solution.

The direct solution of the phonon Boltzmann equation is itself a very difficult task as it is difficult to mathematically express the anharmonic phonon decay process,

doi:10.1088/978-1-6817-4123-9ch4

and in addition one has to solve separate phonon Boltzmann equations for each mode of the acoustic and optical branches. If we now include the electrons and holes in the picture with their corresponding Boltzmann transport equations, then the solution of the electron–hole–phonon coupled set of equations becomes a formidable task, even for today's high performance computing systems. Therefore, some simplifications of this global problem are needed. Since for device simulation we are mainly focused on accurately calculating the I–V characteristics of a device, the self-heating (being a by-product of the current flow through the device) can be treated in a more approximate manner, but which is still more accurate than the local heat conduction model.

In chapter 2 we discussed a self-consistently coupled thermal/ensemble Monte Carlo device simulator that was developed and used to study self-heating effects in different generations of fully-depleted SOI devices. We showed that the pronounced velocity overshoot present in the nanometer scale device structures minimizes the degradation of the device characteristics due to lattice heating. This result does not imply that one does not have to be concerned with heating in nanoscale devices; it only means that we mainly have to focus on efficient ways of removing the heat from the device active region. When examining heating in different device technologies, we observed a bottleneck between the lattice and the optical phonon temperature in the channel, which is more pronounced for shorter devices, due to the fact that the energy transfer between optical and acoustic phonons is relatively slow compared to the electron-optical phonon processes and the fact that the electrons are quasi-ballistic (and since the channel is very short, they spent little time in the channel). To better understand the phonon temperature bottleneck, different cross-sections of the lattice and the optical phonon temperature profiles in the channel direction were investigated. We found that the bottleneck is decreasing from the Si/SiO$_2$ interface towards the Si/BOX interface. For shorter devices, it exists in the whole channel region, which is not the case for longer devices (thicker Si-layer and longer channel length). From the results presented in this book, and those we have published earlier, one can conclude that the higher the temperature in the channel and/or the longer the electrons are in the channel, the larger the degradation of the device electrical characteristics is due to the heating effects.

4.1 The choice of proper thermal boundary conditions

The question of the proper boundary conditions for the electronic part of the problem is rather clear and has been discussed in many papers and texts in the literature. The problem in properly specifying the phonon boundary conditions is the selection of acoustic and optical phonon temperatures at either the artificial boundaries or at the contacts. Typically, acoustic phonon temperature is equated with the lattice temperature. To better understand the choice of boundary conditions that we have investigated in various works [6,7], we want to point out that the lattice temperature is analogous to the electrostatic potential and heat flux is analogous to current. As is well known, when considering the electrical behavior of the device, at least one node within the structure has to have Dirichlet boundary conditions specified. Analogously, for the lattice temperature, we need at least one node that is a thermal contact and whose temperature is set to 300 K. For the case of fully-depleted (FD) silicon-on-insulator (SOI) devices discussed here, the bottom of the box is assumed to be at thermal equilibrium. Also, the gate contact is assumed to be at thermal equilibrium as well in both the structures considered, but not necessarily at 300 K. For example, in one set of simulations for FD SOI devices, we vary the temperature on the gate to be 300 K, 400 K, and 600 K. The next question is: what happens to the source and drain contacts and the side artificial boundaries? Specifying Dirichlet boundary conditions at the ohmic contacts is not accurate from the standpoint that there is current flowing through the contacts, and since the contacts have finite resistance, there will be Joule heating (so the problem becomes unconstrained). The best solution to this, as we have done in several studies, is to extend the metal contact to become part of the simulation domain and to apply at the very ends of the contact isothermal boundary conditions (see figure 4.1). With the substrate and gate electrode boundary conditions specified, we have vertical transport of heat through the structure. The next question is, is there lateral transfer of heat across the artificial boundaries? Here we can consider two cases: the first case when the neighboring device is ON, in which case Neumann boundary conditions are appropriate, and the second case when the neighboring device is OFF in which case Dirichlet boundary conditions are appropriate. When are Dirichlet and when are Neumann boundary conditions the right choice? Let us take for example digital circuits. In digital logic, most of the transistors are idle and there is very high probability that the neighboring transistors are going to be OFF. In that case, applying Dirichlet boundary conditions to the sides is a good choice. Let us now consider analog circuits, the simplest example being a current mirror. In this situation both transistors are ON, so the use of Neumann boundary conditions is appropriate. In the case when Neumann boundary conditions are applied across the artificial boundaries, the heat transport remains vertical, but for the case when Dirichlet boundary conditions are applied across the boundary, the heat transport

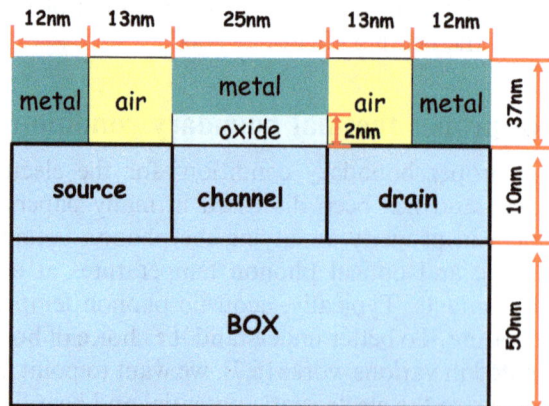

Figure 4.1. Device structure with extended domain. The geometrical dimensions are for the simulated 25 nm channel length FD-SOI nMOSFET.

has both vertical and horizontal components. So, the choice of the best boundary conditions very much depends on the operation regime of the device and whether the device is near a hot spot or not. Proper determination of the temperature of the hot spot requires usage of the thickness and temperature dependent thermal conductivity for thin silicon films.

4.2 Thermal conductivity model currently used in the simulator

Yet another issue that deserves attention in getting physically correct results is the proper choice of the thermal conductivity for thin silicon slabs and for nanowires. Asheghi and co-workers [4] and Shi and co-workers [5] have demonstrated via experimental measurements that the thermal conductivities of thin silicon films and silicon nanowires, respectively, strongly depend on the geometry as for smaller geometries phonon boundary scattering can reduce the thermal conductivity of the silicon film or nanowire by a factor of 10 or more from its bulk value. Moreover, thermal conductivity is a temperature dependent quantity. We have made extensive effort, using the theory of Sondheimer for conductivity of metals [8], to arrive at an empirical formula that simultaneously describes the thickness and temperature dependence of the thermal conductivity. Our empirical expression perfectly matches the experimental data of Asheghi and co-workers [9].

In the model, it is assumed that phonon boundary scattering is perfectly diffusive. The governing equations for the model are given below. First, we use the Sondheimer expression for the electrical conductivity for thin metal films, which is converted into an expression for the thermal conductivity:

$$\kappa(z) = \kappa_0(z) \int_0^{\pi/2} \sin^3 \theta \left\{ 1 - \exp\left(-\frac{a}{2\lambda(T)\cos\theta}\right) \cosh\left(\frac{a - 2z}{2\lambda(T)\cos\theta}\right) \right\} d\theta. \quad (4.1)$$

In the above expression the mean free path is modified according to:

$$\lambda(t) = \lambda_0(300/T), \quad (4.2)$$

where λ_0 is the phonon mean free path at 300 K and the Selberherr's model for the bulk thermal conductivity is used [10], of the form:

$$\kappa_0(T) = \frac{135}{a + bT + cT^2} \text{Wm}^{-1}\text{K}^{-1}, \quad (4.3)$$

where the parameters a, b and c are as follows: $a = 0.03$, $b = 1.56 \times 10^{-3}$ 1/K, and $c = 1.65 \times 10^{-6}$ 1/K^2. A comparison between the experimental thermal conductivity data of Asheghi and co-workers [4] and our analytical model described by equations (4.1)–(4.3) is shown in figure 4.2. Our model results are in excellent agreement with the experimental data of Asheghi.

4.3 Multiscale modeling of device + interconnects

A new way of modeling self-heating in nano-scale devices using multi-scale electro-thermal device simulation is also presented in this book. The solver is applied for

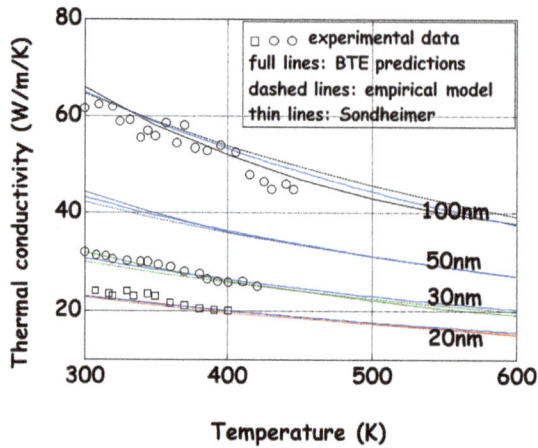

Figure 4.2. Comparison of the experimental thermal conductivity data of Asheghi and co-workers [4] for thin silicon films as a function of the thickness of the silicon film and the temperature.

understanding self-heating effects in 22 nm planar MOSFET devices from IMEC (Belgium). It is important to note that the electron problem is limited to length scales that are on the order of several tenths of nanometers. On the other hand, the phonon mean free path in bulk Si is on the order of 300 nm at room temperature, which is a much larger length scale. Hence, simulations of lattice heating require much larger simulation domains. In addition to this, as semiconductor device scaling progresses towards smaller dimensions, the role of interconnect self-heating has to be accounted for as well. For this purpose, another level of hierarchy in the multi-scale modeling approach is introduced. Giga 3D (Silvaco Atlas module) is used to model the role of interconnects and the role of the larger simulation domain and is also used to extract the temperature boundary conditions for a smaller domain electro-thermal device simulator. The electro-thermal simulator passes Joule heating terms to Giga 3D and the whole Gummel iteration loop is repeated until a self-consistent solution at multiple levels of approximation is achieved.

4.4 Phonon Monte Carlo need and its necessary improvements

To further improve the model accuracy, a phonon Monte Carlo (PMC) based thermal solver is introduced for electro-thermal device simulation. The PMC solver is tested and validated to explain the bulk thermal conductivity of silicon at different temperatures. Comprehensive discussion on different PMC approaches in the literature are presented. Mazumder's approach gives a good match for the thermal conductivity, but the temporal temperature profile is incorrect. The reason is that Mazumder's approach assumes a Bose–Einstein equilibrium distribution for the phonons at the end of each scattering event by choosing $\delta t > 3\tau_{max}$. However, the spectrum of scattering rates shows that a single δt cannot be chosen to fit all the phonons. This is partly avoided in Lacroix's approach where the sampling after the scattering event is done from the modulated equilibrium distribution with probability of scattering. This improves the temporal temperature profile, but the thermal conductivity is not matched with experimental data. To match with experimental thermal conductivity data, the scattering rates are calibrated. After calibration, the experimental data is closely matched with the predicted values.

The calibration is not required for calculating the thermal conductivity of silicon through direct solution methods for the phonon BTE. This shows that modeling of the scattering events in the present approaches is not good enough to explain the thermal conductivity and necessary work has to be done to improve it. One way to improve is to simulate the scattering with energy and net momentum conservation following the full dispersion relation of phonons in silicon. This eliminates the need for sampling the phonons after scattering. This is computationally very expensive and requires more phonons to sample the k-space. Another way to improve is to follow relaxation time statistics without having to ignore any exponential terms in equation (3.48).

References

[1] Raman A, Walker D G and Fisher T S 2003 Non-equilibrium thermal effects in SOI power transistors *Solid-State Electron.* **47** 1265–73

[2] Pop E, Banerjee K, Sverdrup P, Dutton R and Goodson K 2001 Localized heating effects and scaling of sub-0.18 micron CMOS devices *IEDM Techn. Dig.* **679**

[3] Sinha S, Pop E, Dutton R W and Goodson K E 2006 Non-equilibrium phonon distributions in sub-100 nm silicon transistors *Trans. ASME* **128** 638–47

[4] Liu W and Asheghi M 2004 Phonon-boundary scattering in ultra-thin single-crystal silicon layers *Appl. Phys. Lett.* **84** 3819–21

[5] Li D, Wu Y, Kim P, Shi L, Yang P and Majumdar A 2003 Thermal conductivity of individual silicon nanowires *Appl. Phys. Lett.* **83** 2934

[6] Vasileska D, Raleva K and Goodnick S M 2009 Self-heating effects in nano-scale FD SOI devices: the role of the substrate, boundary conditions at various interfaces and the dielectric material type for the box *IEEE Trans. Electron Dev.* **56** 3064–71

[7] Vasileska D, Raleva K and Goodnick S M 2009 Thermal effects in fully-depleted SOI devices *ECS Trans.* **23** 337

[8] Vasileska D, Raleva K and Goodnick S M 2010 Electrothermal studies of FD SOI devices that utilize a new theoretical model for the temperature and thickness dependence of the thermal conductivity *IEEE Trans. Electron Dev.* **57** 726–8

[9] Palankovski V and Selberherr S 2001 Micro materials modeling in MINIMOS-NT *J. Microsystem Tech.* **7** 183–7

[10] Sondheimer E H 1952 The mean free path of electrons in metals *Adv. Phys.* **1** 1–42 reprinted in 2001 *Adv. Phys.* **50** 499–537

Modeling Self-Heating Effects in Nanoscale Devices

K Raleva, A R Shaik, D Vasileska and S M Goodnick

Appendix A

Derivation of energy balance equations for acoustic and optical phonons

The energy conservation equations for phonons are developed as follows. The phonon distribution function $N_k(x, t)$ can be expressed as

$$N_k(x, t) = \langle N_k \rangle_0 + n_k(x, t) \tag{A.1}$$

where $\langle N_k \rangle_0$ is the equilibrium phonon distribution at temperature T_e and $n_k(x, t)$ is the deviation of the phonon distribution function from equilibrium. Some other quantities are defined as the sum of Eigen components as follows:

$$u(x, t) = \frac{1}{V} \sum_k n_k(x, t) \hbar\omega \tag{A.2}$$

$$\mathbf{S}(x, t) = \frac{1}{V} \sum_k n_k(x, t) \hbar\omega \mathbf{v}_k \tag{A.3}$$

$$\mathbf{J}(x, t) = \frac{1}{V} \sum_k n_k(x, t) \hbar\mathbf{k} \tag{A.4}$$

$$t^{ij}(x, t) = \frac{1}{V} \sum_k n_k(x, t) \hbar k^i v_k^j \tag{A.5}$$

where $u(x, t)$ is the energy density, $\mathbf{S}(x, t)$ is the energy flux, $\mathbf{J}(x, t)$ is the momentum density, $t^{ij}(x, t)$ is the momentum flux and $\mathbf{v}_k = d\omega/d\mathbf{k}$.

The collision of phonons with each other and imperfections causes every Eigen component except $\langle n_k \rangle_0$ to decay to zero with its own characteristic relaxation time. In order to apply the relaxation time approximation, a single relaxation time τ is used to characterize the decay to local equilibrium. This can be expressed as

$$\frac{\partial n_k(x, t)}{\partial t} + \mathbf{v}_k \cdot \nabla n_k(x, t) = -\frac{n_k(x, t) - n_k^0(T(x, t))}{\tau} \tag{A.6}$$

doi:10.1088/978-1-6817-4123-9ch5

where $n_k^0(T(x, t))$ is the local equilibrium distribution and the value of $T(x, t)$ is determined according to energy conservation. The phonon energy balance equation is obtained by multiplying phonon BTE with RTA, equation (A.6), by $\hbar\omega_k$ and then summing over all modes as

$$\frac{1}{V}\sum_k \hbar\omega_k\frac{\partial n_k(x, t)}{\partial t} + \frac{1}{V}\sum_k \hbar\omega_k\mathbf{v}_k \cdot \nabla n_k(x, t) = -\frac{1}{V}\sum_k \hbar\omega_k\frac{n_k(x, t) - n_k^0(T(x, t))}{\tau} \quad \text{(A.7)}$$

$$\Downarrow \qquad\qquad\qquad\qquad \Downarrow \qquad\qquad\qquad\qquad \Downarrow$$

$$\frac{\partial u(x, t)}{\partial t} \qquad\qquad\qquad \nabla \cdot \mathbf{S}(x, t) \qquad\qquad\qquad 0$$

Therefore, we have the reduced form of equation (A.7) as

$$\frac{\partial u(x, t)}{\partial t} + \nabla \cdot \mathbf{S}(x, t) = 0. \quad \text{(A.8)}$$

Then again multiplying equation (A.7) with $\hbar\omega_k v_k^i$ where $i = x, y, z$ and summing over all modes gives

$$\frac{1}{V}\sum_k \hbar\omega_k v_k^i\frac{\partial n_k(x, t)}{\partial t} + \frac{1}{V}\sum_k \hbar\omega_k v_k^i\mathbf{v}_k \cdot \nabla n_k(x, t) = -\frac{1}{V}\sum_k \hbar\omega_k v_k^i\frac{n_k(x, t)}{\tau} \quad \text{(A.9)}$$

which reduces to

$$\frac{\partial S_i(x, t)}{\partial t} + \frac{1}{V}\sum_j\sum_k \hbar\omega_k v_k^i v_k^j\frac{\partial n_k(x, t)}{\partial X_j} = -\frac{S_i(x, t)}{\tau}. \quad \text{(A.10)}$$

Introducing the temperature gradient on the second term in the left hand side of equation (A.10) as

$$\frac{\partial n_k(x, t)}{\partial X_j} = \frac{\partial n_k(x, t)}{\partial T}\frac{\partial T}{\partial X_j} \quad \text{(A.11)}$$

we can arrive at

$$\left(\tau\frac{\partial}{\partial t} + 1\right)S_i(x, t) = -\frac{\tau}{V}\sum_j\sum_k \hbar\omega_k v_k^i v_k^j\frac{\partial n_k(x, t)}{\partial T}\frac{\partial T}{\partial X_j}. \quad \text{(A.12)}$$

Let us define κ_{ij} as the thermal conductivity tensor by

$$\kappa_{ij} = \frac{\tau}{V}\sum_j\sum_k \hbar\omega_k v_k^i v_k^j\frac{\partial n_k(x, t)}{\partial T}. \quad \text{(A.13)}$$

Then equation (A.12) can be expressed as

$$\left(\tau\frac{\partial}{\partial t} + 1\right)\mathbf{S}(x, t) = -\kappa\nabla T. \tag{A.14}$$

Applying divergence on both sides of the above equation results in

$$\left(\tau\frac{\partial}{\partial t} + 1\right)\nabla \cdot \mathbf{S}(x, t) = -\nabla \cdot \kappa\nabla T. \tag{A.15}$$

From equation (A.8),

$$\nabla \cdot \mathbf{S}(x, t) = -\frac{\partial u(x, t)}{\partial t}, \tag{A.16}$$

and since $u(x, t)$ can be expressed with heat capacity C_0 we have

$$u(x, t) = C_0[T_0 - T(x, t)]. \tag{A.17}$$

Then, equation (A.15) becomes

$$C_0\left(\tau\frac{\partial}{\partial t} + 1\right)\frac{\partial T(x, t)}{\partial t} = -\nabla \cdot \kappa\nabla T + H \tag{A.18}$$

where the first term on the right hand side means the influx of energy into a volume dV and the second term H, which comes from the time evolution of C_0T_0, denotes the increase in the energy due to electron–phonon interaction in the system. Also, the relaxation time dependent term in the left hand side can be negligible, we have the final form of

$$C_0\frac{\partial T(x, t)}{\partial t} = -\nabla \cdot \kappa\nabla T + H. \tag{A.19}$$

The process in which the energy exchange takes place between the electrons and phonons differs depending on how the particle scattering occurs. Therefore, the energy balance equations are derived separately for acoustic phonons and optical phonons. As the primary path of energy transport is represented first by scattering between the electrons at T_e and optical phonons at T_{LO} and then optical phonons decaying to acoustic phonons at T_A to the lattice at T_L, which is estimated as equivalent to T_A. The energy exchange between the electrons and the phonons for electron-optical energy balance is as follows:

$$C_{LO}\frac{\partial T_{LO}}{\partial t} = -\nabla \cdot \kappa_{LO}\nabla T_{LO} + \frac{1}{2}\frac{\left[3nk_B(T_e - T_{LO}) + m\ast nv_e^2\right]}{\tau_{e-LO}} - C_{LO}\frac{T_{LO} - T_A}{\tau_{LO-A}}. \tag{A.20}$$

The first term in the right hand side goes to 0 because the group velocity of optical phonon is near 0. The second term represents the energy gain from the electrons, and the last term is the energy loss to the acoustic phonons. So the final form is given by

$$C_{LO}\frac{\partial T_{LO}}{\partial t} = \frac{1}{2}\frac{\left[3nk_B(T_e - T_{LO}) + m^*nv_e^2\right]}{\tau_{e-LO}} - C_{LO}\frac{T_{LO} - T_A}{\tau_{LO-A}} \qquad (A.21)$$

The next step of optical-acoustic phonon energy balance is shown as

$$C_A\frac{\partial T_A}{\partial t} = -\nabla \cdot \kappa_A \nabla T_A + \frac{3nk_B}{2}\frac{(T_e - T_L)}{\tau_{e-L}} + C_{LO}\frac{T_{LO} - T_A}{\tau_{LO-A}} \qquad (A.22)$$

where we do not have an electron velocity related term in the second one of the right hand side of course, and if the electron-acoustic phonon scattering is elastic, the whole second term should be excluded. The last term indicates the energy gain from optical phonons is coming from the last term of the right hand side of equation (A.21).